Routledge introductions to environment series

Representing the environment

John R. Gold and George Revill

Routledge
Taylor & Francis Group

LONDON AND NEW YORK

First published 2004
by Routledge
11 New Fetter Lane, London EC4P 4EE

Simultaneously published in the USA and Canada
by Routledge
29 West 35th Street, New York, NY 10001

Routledge is an imprint of the Taylor & Francis Group

Typeset in Times by
Keystroke, Jacaranda Lodge, Wolverhampton
Printed and bound in Great Britain by
T.J. International Ltd, Padstow, Cornwall

British Library Cataloguing in Publication Data
A catalogue record for this book is available from the British Library

Library of Congress Cataloging in Publication Data
Gold, John Robert.
    Representing the environment / John R. Gold & George Revill.
        p. cm. – (Routledge introductions to environment series)
Includes bibliographical references and index.
1. Environmentalism. 2. Environmentalism–History. I. Revill,
George, 1961–   II. Title. III. Series.
GE195 .G65 2004
333.72–dc22   2003026841

ISBN 0–415–14589–9 (hbk)
ISBN 0–415–14590–2 (pbk)

For Denis Cosgrove

Contents

Figures

Tables

Series editor's preface
Environment and
Society titles

The modern environmentalist movement grew hugely in the last third of the twentieth century. It reflected popular and academic concerns about the local and global degradation of the physical environment which was increasingly being documented by scientists (and which is the subject of the companion series to this, *Environmental Science*). However it soon became clear that reversing such degradation was not merely a technical and managerial matter: merely knowing about environmental problems did not of itself guarantee that governments, businesses or individuals would do anything about them. It is now acknowledged that a critical understanding of socio-economic, political and cultural processes and structures is central in understanding environmental problems and establishing environmentally sustainable development. Hence the maturing of environmentalism has been marked by prolific scholarship in the social sciences and humanities, exploring the complexity of society–environment relationships.

Such scholarship has been reflected in a proliferation of associated courses at undergraduate level. Many are taught within the 'modular' or equivalent organisational frameworks which have been widely adopted in higher education. These frameworks offer the advantages of flexible undergraduate programmes, but they also mean that knowledge may become segmented, and student learning pathways may arrange knowledge segments in a variety of sequences – often reflecting the individual requirements and backgrounds of each student rather than more traditional discipline-bound ways of arranging learning.

The volumes in this *Environment and Society* series of textbooks mirror this higher educational context, increasingly encountered in the early twenty-first century. They provide short, topic-centred texts on social science and humanities subjects relevant to contemporary society–environment relations. Their content and approach reflect the fact that each will be read by students from various disciplinary backgrounds, taking in not only social sciences and humanities but others such as physical and natural sciences. Such a readership is not always familiar with the disciplinary background to a topic, neither are readers necessarily going on to develop further their interest in the topic. Additionally, they cannot all automatically be thought of as having reached a similar stage in their studies – they may be first-, second- or third-year students.

The authors and editors of this series are mainly established teachers in higher education. Finding that more traditional integrated environmental studies and specialised texts do not always meet their own students' requirements, they have often had to write course materials more appropriate to the needs of the flexible undergraduate programme. Many of the volumes in this series represent in modified form the fruits of such labours, which all students can now share.

Much of the integrity and distinctiveness of the *Environment and Society* titles derives from their characteristic approach. To achieve the right mix of flexibility, breadth and depth, each volume is designed to create maximum accessibility to readers from a variety of backgrounds and attainment. Each leads into its topic by giving some necessary basic grounding, and leaves it usually by pointing towards areas for further potential development and study. There is introduction to the real-world context of the text's main topic, and to the basic concepts and questions in social sciences/humanities which are most relevant. At the core of the text is some exploration of the main issues. Although limitations are imposed here by the need to retain a book length and format affordable to students, some care is taken to indicate how the themes and issues presented may become more complicated, and to refer to the cognate issues and concepts that would need to be explored to gain deeper understanding. Annotated reading lists, case studies, overview diagrams, summary charts and self-check questions and exercises are among the pedagogic devices which we try to encourage our authors to use, to maximise the 'student friendliness' of these books.

Hence we hope that these concise volumes provide sufficient depth to maintain the interest of students with relevant backgrounds. At the same time, we try to ensure that they sketch out basic concepts and map their territory in a stimulating and approachable way for students to whom the whole area is new. Hopefully, the list of *Environment and Society* titles will provide modular and other students with an unparalleled range of perspectives on society–environment problems: one which should also be useful to students at both postgraduate and pre-higher education levels.

David Pepper
May 2000

Series International Advisory Board

Australasia: Dr P. Curson and Dr P. Mitchell, Macquarie University

North America: Professor L. Lewis, Clark University; Professor L. Rubinoff, Trent University

Europe: Professor P. Glasbergen, University of Utrecht; Professor van Dam-Mieras, Open University, The Netherlands

Preface

Very few books have a single point of origin. In general, they stem from curiosity, idle or otherwise, draw on ideas culled from here and there, are shaped by contact with many different people, and finish up trying the patience of publishing editors. This one is no exception. The human geography modules that we teach at Oxford Brookes University are primarily concerned with urban, rural and industrial landscapes, generally approached from historic and cultural standpoints. Over the years, we have found it extremely difficult to find suitable literature to recommend to students on our courses. What we were ideally looking for were works that combined two things. First, we wanted our students to get an understanding of what representations of environments conveyed by such media as paintings, photographs, newspaper reports, novels, poetry and buildings meant in context. Second, we wanted them to gain some idea about how to approach and analyse these forms of representation. Given that there was nothing available that covered those two simple requirements, we set out in this book to fill that gap ourselves.

In writing it, we have worked to a set of rules, which we formulated by trial-and-error. The five main ones were as follows:

1   We would try to write for our students rather than attempt to impress our peers.
2   We would introduce definitions and contextual material when they are needed rather than have initial theoretical sections followed by case study sections.
3   We would try to work through complex ideas by detailed consideration of examples, along with supporting exercises.
4   We would aim for depth rather than try offering a comprehensive, but inevitably superficial coverage of every form of environmental representation that we could possibly squeeze between the book's covers.
5   Wherever necessary we would be prepared to use exercises, key terms and lists, like this one, to clarify our meaning.

As other textbook writers will know, it is not always easy to stick to those rules, but we have done our best. The book builds explicitly on exercises and case studies that we have used in teaching and which seem successful in communicating ideas and techniques of analysis to our students. We hope that your experience will be the same.

Like all authors, we have a variety of debts that we would like to acknowledge. By the nature of the book, many of those debts are local. Our students, of course, are the ones that have acted as guinea pigs for many of the exercises presented here; sincere thanks to them for their comments and assistance. Next, we would like to thank our colleagues from the Geography Department at Oxford Brookes for their forbearance. Margaret Gold, Kim Hammond and Charles Watkins helped us out with source materials. Our colleague and editor, David Pepper, has been thoughtful and encouraging throughout; indeed we both doubt whether the book would have been finished without him. On a more delicate note, we wish to thank our good friends at Routledge for retaining faith in us. Publishers usually hate authors praising their patience in case it encourages others to behave equally badly, but we would like to record our gratitude to Andrew Mould and Melanie Attridge for tolerance beyond the call of duty. Thanks must also be recorded to the following libraries and their staff for their assistance: the British Library, Senate House (University of London), the British Film Institute, University of Birmingham and Oxford Brookes University. Copyright ownership for illustrations is indicated along with the source. Other items come from the authors' own collections or are photographs taken by the authors. All reasonable efforts have been made to trace whether copyright exists on older items. If anything should have been missed, we would be grateful if the copyright holders would contact us, whereupon we would be pleased to settle matters.

Moving on to family, John would like to convey his love and thanks to Maggie, Iain and Jenny; and George to Eleanor. This, however, should not be seen just as our weak attempts to fob them off for having missed small things like their summer holidays (although an element of that cannot be denied). Rather, the great advantage of having spouses and children 'in the business' is that they willingly supply ideas that have directly shaped parts of this book. Next summer, we promise, will be different. . . .

Finally, this book is dedicated to Denis Cosgrove with thanks for showing us – like so many other people – how to explore this territory.

John R. Gold and George Revill
West Ealing and Newthorpe
September 2003

Acknowledgements

The authors and publishers gratefully acknowledge the following institutions and individuals for giving permission to reproduce illustrations:

Figure 1.1, Tourism Ireland Limited; Figure 2.2, Jill Bray; Figure 2.3, *Guardian*; Figure 2.4, Orange Dog Productions, Sheffield; Figure 2.5, Crown Copyright and the Forestry Commission; Figure 3.1, Board of Trustees of the National Museums and Galleries on Merseyside (Lady Lever Art Gallery, Port Sunlight); Figure 3.2, Friends of the Earth; Figure 3.3, *Private Eye* magazine, Press Association, *The Week* magazine; Figures 4.4, 6.1, 6.2, 6.4, The British Library; Figures 4.5, 4.6, Ashmolean Museum, Oxford; Figures 4.8, 5.3, National Gallery, London; Figures 5.2, 5.8, Yale Center for British Art, Paul Mellon Collection, New Haven, CT; Figure 5.9, Tate Gallery, London; Figure 5.11, Land Rover UK; Figure 5.12, DaimlerChrysler UK; Figure 6.3, National Gallery of Wales, Cardiff; Figure 6.8, Rex Nan Kivell Collection, National Library of Australia; Figure 7.1, Halifax plc.

Every effort has been made to contact copyright holders for their permission to reprint material in this book. The publishers would be grateful to hear from any copyright holder who is not here acknowledged and will undertake to rectify any errors or omissions in future editions of this book.

1 Introduction

This chapter:

- introduces the study of environmental representations;
- explains the structure and organisation of this book.

Contested environments

Few developments in contemporary thought owe more to the power of imagery than the emergence of the modern environmental movement. From the early days of that movement in the 1960s, environmentalists clearly understood the importance of finding powerful *images* to represent ideas that otherwise might be difficult to grasp. Few were more effective in this respect than the American genetic biologist Rachel Carson (1907–64).

Exercise 1.1

The extract below comes from Rachel Carson's book *Silent Spring* (1963), widely regarded as a founding work of modern environmentalism. This passage comes from the first chapter of the book, which is entitled 'A Fable for Tomorrow'.

Summarise the case that Carson is making.

> There was a strange stillness. The birds, for example – where had they gone? Many people spoke of them, puzzled and disturbed. The feeding stations in the backyards were deserted. The few birds seen anywhere were moribund; they trembled violently and could not fly. It was a spring without voices . . .

> On the farms the hens brooded, but no chicks hatched. The farmers complained that they were unable to raise any pigs – the litters were small and the young survived only a few days. The apple trees were coming into bloom but no bees droned among the blossoms, so there was no pollination and there would be no fruit.

> The roadsides, once so attractive, were now lined with brown and withered vegetation as though swept by fire. These, too, were silent, deserted by all living things . . .

> In the gutters and under the eaves and between the shingles of the roofs, a white granular powder still showed a few patches; some weeks before it had fallen like snow upon the roofs and the lawns, the fields and streams.

> No witchcraft, no enemy action had silenced the rebirth of new life in this stricken world. The people had done it themselves.

>   (Carson, 1963, 22)

Your answer might highlight the following:

- The silence of the countryside is due to elimination of much of the wildlife.
- The flora as well as the fauna seems to have suffered, with the suggestion that the entire rural ecosystem had broken down.
- The cause lies in the granular white powder that fell from the sky some weeks earlier, although it is unclear from this extract whether it was deliberately dropped, as in aerial spraying, or whether it was an accident.
- The idea that this is 'a fable for tomorrow' suggests that this might be a warning, of the type found in science-fiction literature, rather than description of an actual event.

The key to understanding this extract lies in knowing that the granular white powder was agricultural pesticide and that, although Rachel Carson was describing a hypothetical situation, all the individual elements had 'actually happened somewhere' (Carson, 1963, 22). Carson had become increasingly concerned about the arbitrary use of farm pesticides, particularly DDT (dichlorodiphenyltrichloroethane), for their indiscriminate effects on wildlife. In the late 1950s, several American states had used DDT in widespread aerial spraying programmes aimed at pest control. Her book, which had the original working title *The Control of Nature*, appeared against the background of that programme and other instances of pesticide misuse (Payne, 1996, 142). Yet rather than being a straightforward academic counter-argument to the case for using agricultural pesticides, Carson created a powerful and easily understood vision of the potentially devastating consequences of uncontrolled usage. In the words of Stewart Udall, US Secretary of the Interior between 1961 and 1968, *Silent Spring* was 'an ecology primer for millions' and played an inestimable part in 'the ecological reawakening of America' (Udall, 1963, 137; Payne, 1996, 137).

Other early environmentalists followed similar strategies, also finding readily accessible images to convey complex ideas. One technique harnessed environmental arguments to popular themes of the day, such as space travel, concern about nuclear weapons and anxieties about population growth. Kenneth Boulding (1966), for instance, likened the earth to a single spaceship, in which resources needed constant recycling. Paul Ehrlich (1968) compared explosive population growth to a bomb possessing the capability to destroy existing patterns of life. Garrett Hardin (1968) put forward a parable of the behaviour of herdsmen tending cattle on common pastures as the basis for presenting more general principles about demographic increase and resource use. Although later criticised for being simplistic or misleading, these ideas were undeniably important for crystallising debate around easily grasped images.

The same strategy of introducing complex ideas by using striking images continues to this day, particularly by employing powerful visual images. For example, advertising campaigns by environmental groups have included the following photographic images:

- Factory chimneys belching out plumes of smoke.
- Industrial effluent spewing into rivers from corroded pipes.
- Trees dying from the effects of acid rain.
- Cars snarled up on urban motorways and emitting exhaust fumes.
- Unarmed protesters in small, vulnerable vessels confronting huge whaling vessels.
- Seal pups on the ice floes being clubbed to death by hunters.
- Oil-coated seabirds floundering on blackened beaches.
- Children staring at the carcass of a dead dolphin, washed up on a beach after an incident involving toxic chemicals.

The subject in question might be, among other things, global warming, environmental pollution, public health, caring for wildlife, the costs of ecological disaster or the case for public transport. Whatever the specific issue, the frequency with which such images appear certainly suggests that those who deploy them believe that they are effective in getting their message across.

Yet all such images are selective *representations* of complex issues. Those who design them actively campaign to awaken public consciousness over misuse of the environment and shape their communications to create and reinforce that message. Moreover, the exchange is not one way, since many actively contest the views put forward by environmentalists. For their part, corporate and industrial interests quickly learned that they too could employ powerful imagery to counter, sometimes pre-emptively, the claims of the environmental lobby (e.g. Wilson, 1992; Anderson, 1997). Industrial corporations in the energy, oil, tobacco and chemical industries, for example, routinely monitor current environmental debate. Many employ entire departments to take responsibility for relations with the press and broadcasting media, devising favourable materials for distribution. They sponsor 'think tanks' or pressure groups sympathetic to their interests, which produce seemingly independent reports and other materials aimed at influencing the public or politicians (Beder, 1997). They also commission corporate 'green' advertising, in which they present their activities as vigorously promoting a better environment. In turn, their readiness to promote their case leads environmental lobbyists, within the limits of their budgets, to employ professional agencies in the battle for public opinion. What was once an area of informal, even amateurish, communication on the fringes of mainstream political debate, now increasingly sees the involvement of media consultants and specialist agencies. Environmental debate has become a battleground where the contestants pitch their contrasting and equally selective representations of environmental problems at one another.

Environmental representations

So far, we have talked about the 'environment' in connection with environmental protest and activism. Yet while a convenient way to introduce the subject, these activities only represent the tip of an iceberg. Each day the media bombard us with countless images of familiar and less familiar environments. Billboard posters, property advertisements, food packaging, radio broadcasts, T-shirts and picture postcards, among many others, can all convey ideas and images of the world around us. Sometimes, the environments depicted are clearly chosen to illustrate the theme of a news item or story. Television journalists, for example, often position themselves in front of graffiti-covered walls or crumbling tower blocks when presenting items on life in the inner city. Newspapers carry pictures of semi-submerged homes and cars to make tangible more abstract discussion about, say, flood control or the long-term impact of global warming. Documentaries on economic change frame shots of landscapes scarred by the abandoned machinery and mineral workings of a previous age of industrialisation.

At other times, what appear to be incidental depictions of specific environments turn out to be conscious strategy. The film industry supplies many good examples. Directors of gangster and science-fiction movies commonly use dark and brooding urban-industrial sets to enhance their film's message. Such cities are characteristically portrayed as overcrowded, claustrophobic, dark and violent; a place where good struggles to overcome inherent evil. The makers of period costume dramas often choose locations that feature cobbled city streets, which are either studio sets or real-world sites carefully screened to remove any tell-tale signs of modern times. Films with romantic story-lines are located in small towns surrounded by idyllic countryside, thereby linking the film's content to images of pastoral tranquillity and perceived social stability. Similarly, films depicting the life and times of the extremely wealthy choose environments that testify to social exclusiveness, such as landed estates with their parklands and manicured lawns. In each case, choosing appropriate visual settings is vital for the plausibility of the action.

Contemporary product advertising shows similar sensitivities. Advertisers recognise that suitable environmental associations can enhance their selling message. For example, although now having a product often subject to official disapproval, cigarette advertisers work hard to associate their products with, say, the deserted Western landscapes of Marlboro Country or bustling café society (Virginia Slims). An advertising campaign for the British do-it-yourself chain Homebase showed a kitchen interior with the Eiffel Tower glimpsed from its window, attempting to counter the advertiser's reputation for budget-conscious furnishings and fitments with suggestions of French chic and sophistication. Elizabeth Arden cosmetic advertisements find elegant women applying their make-up against a background of shops on New York's Fifth Avenue or the Manhattan skyline. Food advertisers choose Alpine meadows as the backdrop for advertising Swiss cheese ('Gruyère: the natural choice') or a timeless Italian hilltop village as the setting to advertise Italian dairy products ('Food from Italy: the quality of life'). Compact saloon cars designed for town driving appear against the sophisticated surroundings of modern detached villas ('Volvo for Life') or elegant town houses ('Renault Avantime: there is no such thing as a casual observer'). Luxury four-wheel drive motor vehicles (see also Chapter 5) are set against the rugged grandeur of the Colorado Rockies (Jeep: 'Don't compromise. Anywhere') or crossing a river in an African game reserve in the company of exotic wildlife (the 'Land Rover Experience').

We take the content of these advertisements for granted, often scarcely giving them a second glance when browsing through a newspaper or journal. Yet, when we know how to interpret them, they reveal much about the *values* of those who designed them and, more generally, about the meanings that places and landscapes hold.

Exercise 1.2

Look at the advertisement shown in Figure 1.1. It was placed by Holidays Ireland, a tourist agency, in May 1997. It appeared on the front page of the *Independent*, a British broadsheet daily newspaper catering for a predominantly middle-class readership.

What message do you think the advertisers are trying to put across to the newspaper's readers?

Figure 1.1 *Holidays Ireland advertisement, May 1997. Courtesy of Tourism Ireland Ltd*

The answer that you have reached depends on what you are looking for, on your previous knowledge and on the information that you have available. Although seemingly uncomplicated, this advertisement carries a surprising variety of meanings. It can be interpreted:

1   In terms of *photographic technique*. There are many ways to take a photograph and professional photographers are trained to use their skills to get the

result that they want. The person responsible for this picture will have made choices about the composition, the angle of regard of the camera, the lens to use, and the exposure. As this picture was taken using conventional film rather than digital technology, the photographer also chose the appropriate film and probably also the methods for developing it. Each of these elements can make a difference to the final picture. Looking at Figure 1.1, we can infer that the monochrome photograph that occupies most of the advertisement was taken with a wide-angle lens from a high vantage point. Reproduced in portrait format (where the vertical dimension is greater than the horizontal), it offers considerable depth of field and sharp focus from foreground to the far distance. These techniques, which some might call 'tricks of the trade', are ways of giving a three-dimensional appearance to a scene that by definition is rendered in a two-dimensional form (a photograph on a flat page).

2   An example of *pictorial composition* frequently used in promoting Irish tourism. It illustrates the sunlit summer face of the sparsely populated and rocky West Coast of Ireland. The single car conveys a sense of solitude, with the road winding into the middle-distance drawing the eye to the 'unspoiled' scenic splendours that lie ahead. In case these landscapes should seem un-appealingly isolated, the caption stresses that the experience of seclusion is combined with liveliness, hospitality and the 'bubbling laughter of people keen to embrace the visitor'. Yet returning to the photograph, it is worth remembering that the picture that we see will have been chosen after lengthy decision-making. A photographer on location invariably takes many pictures, experimenting with different locations and perspectives, with combinations of near and far objects, as well as with light, focus, film and lenses. The photographer then supplies a set of photographs for consideration and may, or may not, be party to the final decision as to which to select. Indeed the graphic designers (who designed the advertisement), or the advertising agency (that placed the advertisement), or senior executives at Irish Holidays (who commissioned the campaign) or any combination of them, may have made the final choice.

3   As part of a *media strategy*. Advertising is normally linked to campaigns that last for a certain length of time, based around a particular selling message and using a specific combination of media. This advertisement was one of a number placed by Holidays Ireland in a campaign that spanned the press, posters and television. Placed on the front page of a British national newspaper, where space is expensive per column inch, it carried a more limited content than other, larger advertisements in the series. As such, it served the function of 'reminder' advertising, a cost-effective method by which small advertisements are placed frequently in prominent locations to reiterate a central theme of an ongoing campaign to a target audience.

4   As a *political statement*, since the promotion of Irish tourism is now in the hands of an all-Ireland agency. Dating from early 1997, this advertisement represented part of the first campaign to treat the whole island as a single holiday destination; indeed the first time anywhere that the two countries had run a joint advertising campaign. Other advertisements in the 'Live a Different Life' campaign featured scenes from north as well as south of the border, such as the Devil's Causeway on the Antrim coast. An important aspect in this respect is the logo shown in the bottom-left corner. Created by Design Works, a Dublin-based studio, it depicted, in a stylised manner, two people embracing and exchanging a tiny shamrock – the symbol of Ireland.

5   As a *promotional statement*, especially as a way to counter negative messages about Ireland. The still lingering threat of the revival of bombing campaigns, at least by nationalist splinter groups, and the long and difficult history of relations between Britain and Ireland undoubtedly deterred some potential British visitors. The advertisement posed an image of an Ireland that was at once warm, open and welcoming.

These five perspectives by no means exhaust the possible interpretations of this advertisement, but serve to show that even apparently straightforward depictions of ordinary landscapes can conceal extraordinary complexity of meaning. Our aim in this book is to provide guidance through this complexity, exploring the ways in which significant contemporary and historic representations of the environment were produced, communicated and consumed within society.

Representation, values and culture

It will be said on various occasions in this book that a key term has a long and complex pattern of usage – definitional complexity is, indeed, an occupational hazard in cultural studies. Three words that we need to introduce at this point certainly have that character.

● 'Representation'. The *Oxford English Dictionary* lists eight major uses for the term, the earliest dating back to the early fifteenth century. A 'representation' is broadly defined as an image, likeness or reproduction of something. The reproduction is normally in some material or tangible form, such as a painting, a drawing or a photograph, and may well have a symbolic content. Representation also means to speak or campaign for something else; an activity that inevitably involves holding and acting upon certain sets of ideas and values as opposed to others. In this sense, then, an essential aspect of representation lies in attempting to convince others about the validity of some point of view.

- The word 'values', which arises in that definition, is associated with ideas of worth. Values are enduring beliefs that particular courses of action, or states of affairs or sets of ideas are preferable to other possible courses of action, or states of affairs or sets of ideas. They are important aspects of the way that individuals and groups cope with complexity, because they predispose us to things that we encounter in particular ways.
- Values themselves are an important part of 'culture'. Culture was originally connected with 'cultivation', in both the agricultural sense of tillage and the philosophical sense of the training or improvement of the faculties. It still has those overtones, in that someone possessing culture has breeding, class or worth. People talk about 'high culture', as expressed by great literature, poetry, classical music, art and philosophy. Culture, however, is not restricted to the realm of the fine arts. In this text, we also use the word 'culture' to describe *any aspect of social exchange that communicates attitudes, values and opinions.* Viewed in this way (see Table 1.1), culture provides a framework in which to group all manner of behaviours, objects, media, art and ideas without any implication of intrinsic social approval. Depending on the topic being investigated, we can regard the products of newspapers, film, television and radio as equally worthy of study as paintings, sculpture or formal landscape gardens.

Taking the above three definitions together, we now see that representation is clearly a cultural process. Bearing in mind too that 'representation' also means to speak or campaign for something else, then we may say that it is also a *political* process. To elaborate, one function of culture is that it is the framework through which the real world is experienced, and intrinsic to that framework are ideas about the social order and about who possesses power. We deal with such issues further in Chapter 3, but in the meantime it is worth noting that any representation can be viewed as much as an expression of power as of culture. It is something that contains important messages about those who produced it and about the interests of the people for whom the representation was being designed. When read carefully, a representation may say much about whose views and values are accepted as valid or true, and whose interests are marginalised or even simply not represented.

To get further insight into these general definitions, it is worth returning to consider the example of the advertisement shown in Figure 1.1. It contains particular

Table 1.1 *The scope of culture*

Cultural expression	Product
Behaviour	Dress
	Etiquette
	Hobbies
	Leisure interests
	Sport
Everyday objects	Buildings
	Cars
	Clothes and fashion
	Domestic goods
	Structures
Media	Film and video
	Internet
	Posters
	Print
	Radio
	Television
Art	Architecture
	Cinema
	Literature
	Music
	Painting
	Poetry
	Sculpture
	Theatre
Ideas	Philosophies
	Political ideologies
	Religions

representations of Ireland and Irish life through a photograph that reproduces a scene from the West of Ireland accompanied by text that conveys aspects of Irish life. Both the photograph and the text are highly selective in terms of the view of Ireland that they convey. Part of the reason for that selectivity stems from the values of those professionals who produced this advertisement and those executives who commissioned the campaign. The view depicted in the photograph, for instance, has been selected because it is believed that the audience will consider it beautiful. What comprises beauty, however, depends on the culture and on the period under consideration. As Chapters 4 and 5 show, there have been times when wild and lonely places such as this one would not have been regarded as beautiful, but rather as terrible and desolate. Present-day society, however, tends to treasure the seclusion of such places, for example, as a contrast with everyday life and a 'chance to get away from it all'. Those who have designed this advertisement appreciate that and have tried to build on the selling potential of this culturally valued environment.

The role of this advertisement as part of a *political* process is also apparent. First, as noted above, this advertisement had a context in its specific relationship to arrangements established as part of the Good Friday Peace Agreement. Its emphasis on the quiet tranquillity and friendliness of the island of Ireland conveys impressions that fit that purpose. Second, and related, it represents the interests of those groups in society that stand to gain from expansion of particular forms of tourism as against those groups who have little or nothing to gain. Finally, and more generally, like most advertisements, its aim is to persuade. By its content and the medium in which it was published, it targets particular groups of people as potential visitors and, at least implicitly, seeks to present the advantages of Irish holidays against alternatives.

Seen against this background, we understand *environmental representations* in this book as *written and visual depictions that reproduce or resemble real or imagined places, regions and landscapes*. They are essential elements of the cultures in which we live and, as later chapters show, can be found in media as diverse as tourist literature, poetry, television, product advertising, science-fiction cinema, paintings and drawings. At the outset we can note that, whatever the type of environmental representation under consideration, it is essential to recognise the selectivity that is present. Choosing to show or emphasise a particular aspect of the environment may well mean downplaying or ignoring others. Such selectivity may occur without much forethought, with those who produce representations primarily replicating ideas and images that are ingredients of the culture of which they are part. At other times, the message is deliberately shaped to put forward a specific case.

The study of environmental representations has a recognisable appeal to researchers since it provides a window on many underlying issues. There is an inherent fascination in seeing how people in different societies or at different times have thought about and depicted the environments around them. Studying environmental representations sheds light on the relationships between people and their environments. Among other things, it provides insight into:

● The nature of environmental values; for example, how and why different societies venerate some places and abhor others.

- The reasons why certain sets of values come to dominate decision-making processes while others are rejected.
- The links between environmental representations and issues involving power, authority, ethics and social justice.

To reveal these elements requires careful research, however, since it is dangerously easy to reach simplistic conclusions about, say, the values of those who produce environmental representations or about the relationships between their producers and audience (or consumers). One way to reduce such problems is to focus in detail on specific, but important, aspects of environmental representations rather than attempt to provide a comprehensive, but inevitably superficial coverage. In this book, therefore, we have chosen to study two significant areas in depth, namely: representations arising from changing notions of nature and landscape; and representations of the urban environment. Another way to reduce problems of drawing simplistic conclusions is to pay close attention to the methods involved in assembling and interpreting data. We therefore proceed on the basis that concern for appropriate methodology is a cornerstone of effective interpretation. Throughout, this text takes practical examples to illustrate how representations can be used to explore issues relating to environmental values and actions. In the process, we seek to reveal the pitfalls as well as the benefits of studying environmental representations.

Organisation and structure

Following on from these points, this book is in three parts:

1   Chapters 2 and 3 deal with the methods and techniques necessary to study environmental representations, using a series of historical and contemporary examples. Chapter 2 opens by considering approaches necessary for reading environmental representations. Using examples drawn from instances of road protests and tourist literature, it examines different forms of cultural representation and explores key ideas about culture and representation. While this chapter is primarily concerned with the structure and form of representations, Chapter 3 examines their social context. It explores strategies for studying the audience's consumption of environmental representations, considers the contexts in which representations are produced and considers the way that meaning is actively constructed. After introducing a series of key concepts, it outlines the basic approach to environmental representations adopted in subsequent chapters.

2   Chapters 4 to 6 provide detailed analysis of environmental representations arising from Western society's changing notions of nature and landscape and

the way that people over time have actively constructed representations of the world around them. The three chapters here are arranged in chronological sequence. Chapter 4 traces the origins of Western environmental values in classical, medieval and Renaissance approaches to nature and landscape. Chapter 5 then traces the development of approaches to nature and landscape during the Enlightenment and the period marked by Romanticism. Chapter 6 concludes this section by considering imperialist approaches to nature and landscape, emphasising the relationship between empire, exploitation and control.

3   Chapters 7 and 8 examine representations of urban environments. Chapter 7 notes the values, sympathetic and hostile, that have pervaded representations of the city since classical times. It charts the reasons for the dominance of hostility found in representations of the city in Western society since the spread of industrialisation. After looking at contrasting representations of neighbourhoods within the American city, it examines issues arising from contemporary efforts by city authorities to use publicity and marketing techniques to promote their cities. Chapter 8 extends the analysis of values by considering representations of historic cities, with particular reference to landscapes of power and the narratives associated with heritage. We then examine representations of cities found in two very different portrayals of future urban environments, examining the contrasting views of science-fiction film and the literature on urban sustainability.

The conclusion, Chapter 9, briefly reviews the key themes discussed, identifies enduring ideas of both historical and contemporary importance in environmental representation and examines some of the strategies open to us when we represent the environment.

2 Studying environmental representations

This chapter:

- highlights different forms of cultural representation;
- explores key ideas concerning representation and culture;
- considers the ways in which representations are structured as systems of signs and symbols;
- outlines some basic techniques and approaches necessary for reading environmental representations;
- introduces the question of the audience's consumption of environmental representations.

Finding convincing ways of studying the taken-for-granted features of everyday life is often perplexingly difficult. It is often hard to take the familiar routines of our lives seriously as subjects for sustained analysis without a feeling of drastically overcomplicating simple things. After all, if people deal with something daily without much apparent thought, why is it necessary to be more precise about the elements, relationships and processes involved? For environmental representations, as with many seemingly mundane matters, we would argue that much can be gained from treating them in a more conscious, reflexive and systematic way. With these thoughts in mind, this chapter surveys ideas that assist an understanding of the forms that environmental representations can take and outlines methodological issues arising when studying them. The later sections of the chapter consider how environmental representations are received, or consumed, by their audience, arguing that representations require active participation on the part of the reader or audience if they are to convey their meaning.

Representing road protest

Representations of road protest, a prominent focus for environmental debate in the United Kingdom during the 1990s, provide a useful point of departure for our discussion. The road protest movement fiercely criticised increasing dependence on private road transport and mounted sustained opposition to new developments. Television and newspaper reporting of the activities of road protesters gave extensive coverage to non-violent, and occasionally violent, confrontations between protestors, police, officials and developers. As a result, verbal and visual representations of the anti-road movement disseminated widely through society. The term 'eco-warriors' entered common currency as a description for those involved in civil disobedience in support of environmental causes. Coverage of protests commonly conjured up images of two sets of entrenched protagonists. On the one side were hirsute and colourfully dressed protesters living in camps made from temporary shelters (benders), maintaining treetop vigils and placing themselves between the development site and bulldozers. On the other side were clearance machinery, contractors and security staff in hard hats and yellow jackets, with an accompanying police presence. The contested terrain comprised barbed wire fences, posted eviction orders and swathes of open ground bulldozed through the trees.

Figures 2.1–2.4 offer four different representations of road protest relating to two major developments: the Newbury by-pass in Berkshire (1992) and the extension of the M3 motorway over Twyford Down near Winchester in Hampshire (1993). The British media reported both schemes prominently. Each was central to government road building plans in southern England and the attacks on their construction by protest groups drew a vigorous response from the authorities. There were also many themes linking the two schemes. Many individual protestors and protest groups participated in protests at both Newbury and Twyford Down, as well as other road construction schemes projected at the time. The material presented in Figures 2.1–2.4 represents a selection of statements sympathetic to the cause of the road protest movement and deals with the concerns and struggles experienced by protestors as they sought to prevent or, at least, delay construction. Three directly record statements made by active road protestors; the other reports on the actions of protestors and road builders. Yet despite apparently taking the same side, these four representations differ considerably in the ways that they present their message and treat the underlying issues.

Exercise 2.1

This exercise asks you to think about the meaning of the four representations shown in Figures 2.1–2.4. Study each of them in turn and list the similarities and differences between these four ways of representing road protests.

The following questions are helpful in approaching this task:

- Which form of representation do you find most easily understandable?
- Which representation best conveys factual information and which best conveys thoughts and emotions?
- Which seems most personal and individual?
- Which seems most realistic?
- Which most truthful?
- For whom and why are these representations produced?
- How are they consumed?

Learning to climb had become essential. The treehouse in the tall tree had been empty for nearly two weeks, the north tree even longer, ever since Adey went to Castle Wood. I liked the idea of defending the tall tree, it was a classic Swampy treehouse, way up high in the top twigs, but I'm not used to climbing enough. I'd moved into the north tree. Zwee who'd come down permanently now, moved into the tall tree.

Unfortunately, the north tree really only had room for 1½ people to sleep comfortably. Helen and I did stay here for a couple of weeks, but one of us wouldn't sleep well, usually her. Weekends, with no threat of eviction in the morning, were a welcome chance to sleep on the ground and sprawl out in comfort. In the final weeks when we were really reliant on one another's strength, we'd go to bed together on the ground, and I would go up the tree when I really couldn't stay awake any more.

The first eviction was at Snelsmore. They came at 4.30am and worked all day. By the end of the day, two people were down from the trees, not one tree had been captured. The police tried to keep a cordon round the trees that night, but we knew the terrain and they didn't, so they withdrew to just guarding the tunnels. It felt like a victory.

Figure 2.1 *Autobiographical testimony of Merrick. Here road protestor 'Merrick' describes life at Mary Hare's field during the environmental protest known as the 'Battle of the Trees', which accompanied the construction of the Newbury by-pass in 1996*

Source: Merrick (1996, 58–9)

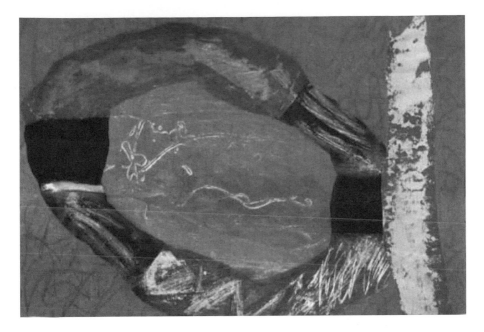

Figure 2.2 Passing the Chalk *(1992) by Jill Bray*

THE GUARDIAN
Saturday April 17 1993

Inquiry 'was misled' on M3 at Twyford

John Vidal

THE Government was accused yesterday of having deliberately misled the public into accepting the M3 extension through Twyford Down in Hampshire.

The pressure group Friends of Twyford Down claims that Department of Transport photomontages produced at the 1985 87 public inquiry into the proposed extension were up to 56 per cent inaccurate, suggesting a cutting up to 150ft narrower than the one now being driven through the chalk hill outside Winchester.

"No one could have been aware that the cutting would be so severe on the landscape," said solicitor Elizabeth Loughran. She will use calculations by a physicist in a legal submission to the European Commission that the Government "substantially" misled the public on the road's full environmental impact. "The photo montage was the only way, apart from an inaccurate model which showed the cutting as a small indent in the landscape, that the public could appreciate the road's visual impact," she said.

Dr Christopher Gillham, a member of the pressure group, has calculated from computer studies, maps and comparative photographs that the Govern-

ment under represented the dimensions of the cutting by up to 56 per cent.

"But it doesn't take arithmetic to see how much the public were misled," he said. "It is obvious to the eye."

At the inquiry the department said the cutting would be 350ft wide and 98ft deep but later admitted that it had not included on the photo montage two four metre-wide "avalanche belts" alongside the carriageways. Dr Gillham said this would make only a 7 per cent difference in its size.

A department spokesman said yesterday: "The technical drawings were all displayed at the inquiry. The montages were as accurate as we could have made them, but it's difficult to give more than an impression. At the end of the day what counts are the technical drawings."

Ms Loughran said: "We are asking the European Commission to reopen its Inquiry into the environmental consequences of the M3 through Twyford Down."

The EC, she believes, could still refer the case to the European Court of Justice if it considers that the Government infringed its directives.

The cutting will destroy two government designated Sites of Special Scientific Interest and two scheduled ancient monuments in an Area of Outstanding Natural Beauty.

Cutting edge . . . Work continuing yesterday on the disputed excavation through Twyford Down, Hampshire, for the motorway extension PHOTOGRAPH ALEX MACNAUGHTON

Figure 2.3 *Newspaper report by John Vidal,* Guardian, *17 April 1993.* © Guardian

YOUR BLACK HAT

Come on then
you yellow bellied scum
Get your newly polished
Black leather boots on.
Shining with hate,
While we wait here
in the darkness
Flinching at shadows
Tying our brains in knots

The power of this place
Gives us what we need
To fight you with open arms
And full hearts
To the sweet end of our part. In all this
aliveness.
We hold the Spring sap
of these trees in our feet
And you can't touch it.

So when you're done
And your black boots are
muddied with innocent blood
And you think we're crushed and homeless
And your cock is hard
With desire
For our pain
And your power

Think again,

All of this will grow elsewhere,
And it won't forget your fat face
Or the number on
Your black hat.

Figure 2.4 *'Your Black Hat' by Hattie. Poem from Kate Evans' book* Copse: a cartoon book of tree protesting. *This poem was written by Hattie, a protestor living at the Fairmile Camp on the route of the proposed Newbury by-pass*

Source: Evans (1998, 139); courtesy of Orange Dog Productions, Sheffield

Discussion

A cursory glance at the contents of Figures 2.1–2.4 reveals their contrasting contents. Taking them in order, the autobiographical account by the road protester Merrick (Figure 2.1) has the following features:

- It is written in an accessible conversational style in a narrative (story-like) form.
- It supplies a chronicle of events on-site from an insider's perspective.
- It contains a combination of reflective commentary, diary excerpts and specialist knowledge of people, places and techniques that are assembled to support Merrick's claim to describe these events from the perspective of the protestors.
- It tells us about the day-to-day experience of being a road protestor, recounting the practical difficulties that the protestors faced and how they overcame them.

The next representation, the painting by Jill Bray (Figure 2.2), has a very different character, in that:

- It adopts an abstract, yet also very personal means of expression, to depict the road protest movement.
- It conveys the emotional response of a particular road protestor.
- Its meaning is very difficult to discern without additional information, such as a carefully chosen title that helps to fix the meaning of the image, and without some background knowledge of the language and vocabulary of the specific mode of representation (abstract art).
- For many people, this may never seem more than a reproduction of an arrangement of shapes and colours on a canvas.

The report from the *Guardian* (Figure 2.3):

- Uses a combination of words and pictures to convey meaning.
- Adopts an ostensibly dispassionate documentary style and includes factual material on the size of the development and nature of the planning exercise alongside opinions and comments by protestors.
- Reinforces the ideas expressed in the written content with a carefully chosen photograph.
- Gives an informed outsider's perspective on the unfolding events of the road's construction.

The poem by the road protestor Hattie (Figure 2.4):

- Contains passionate and uncompromising statements, charged with colloquial-isms and sexual allusions, that convey the impression of heart-felt sentiments and personal commitment.

- Communicates complex ideas in short phrases that create meaning through the associations they raise with other words and thoughts.
- Is very personal in its means of expression, although use of the English language rather than abstract art makes its meaning more immediately accessible than the painting by Jill Bray.
- Possesses layers of meaning that emerge only with detailed attention to structure and vocabulary rather than from an initial reading.

With these points in mind, we now return to the questions set in Exercise 2.1. There is little doubt that Figures 2.1 and 2.3 are the most easily understood examples, since they are written in everyday English and in familiar prose styles. These are also the examples likely to be considered best at conveying factual information, whereas those presented in Figures 2.2 and 2.4 seem more personal and better vehicles for imparting emotion. Judgements become less straightforward when considering issues of realism and truth. It is then necessary not only to think about how representations convey their meanings, but also about the highly contentious issues of what is meant by 'real' and 'true'. Regarding realism, it might be argued that Figures 2.1 and 2.3 are the more realistic since the form of their written and photographic content more directly reflects the world as we know it. Nevertheless, *even these modes of representation are carefully composed*. Editors, for example, select a picture circumspectly to complement the meaning of a particular story. Just as with the Irish holiday advertisement in Chapter 1, the photograph itself was not a random snapshot but a composition produced by a trained professional photographer and designed to tell a story – the congested road snaking round to the left, the construction work in the middle distance, the scarred chalk white of the ravaged hillside. Similarly, Merrick skilfully combined commentary and diary sections. This gave his writing realism as the testimony of a first-hand witness and credence as the expression of a reflective and rational observer. Yet even the most naturalistic and readily understandable forms of representation are actively produced cultural constructions. In terms of 'truth', the question is more difficult still, because it depends on what type of truth we are prepared to admit. Truth can be a mere chronicle of events verified by concrete evidence or the expression of a profound personal emotion that is impossible to pin down. In their own ways, these four representations express different kinds of truth.

To understand more about this matter and the implications of specific meanings, however, it is essential to know more about each representation and the circumstances of its production. Published in a book, Merrick's account was clearly written with an audience in mind, yet its eventual audience was circumscribed by its availability. As a text produced by a radical publishing house and sold through specialist and alternative booksellers, it would only attract a small readership. (Similar accounts have now been published on the Internet, both through the difficulty of finding a publisher and as a political statement by offering a form of publishing that is free to users and minimises resource use.) Merrick's style reflected his expectations of that audience. It was both intimate and confessional yet always with an eye to making the case for the protestors and describing their trials, tribulations and motives. He wrote for like-minded individuals while, at the same time, trying to set the record straight for posterity – at least, the record as he saw it.

The newspaper account was almost certainly the most widely disseminated of the representations reviewed here. The *Guardian* is a major British national broadsheet newspaper, with a circulation of around 385,000. Its editorial policy favours left-of-centre politics, with the term 'Guardian reader' associated with a segment of well qualified middle-class professionals often employed in government, health, education or welfare-based occupations. Against that background, it is not surprising that the paper adopted a line sympathetic to the protest. Its readers are probably favourably disposed towards environmental causes, with many passively supporting environmental activism. Nevertheless, as a respected national newspaper, there are conventions of journalistic impartiality to be observed. The reporter attempted to balance statements from the two sides and to provide context in which to situate the dispute, even if the underlying message was sympathetic to the activists' cause.

Unlike a book or newspaper, a painting is a single image hanging on a wall in a specific place. It will only be widely viewed if it is displayed in an art gallery, or is part of a touring exhibition, or reproduced as a photograph in a book, postcard or poster. In the past this was not necessarily the case since paintings occasionally played a more significant role in community life. Public art exhibitions, for example, once assumed greater importance as social events. In addition, churches commissioned and displayed art on religious subjects, thereby playing a central role in education and worship. By contrast, gallery visits now primarily appeal to particular socio-economic groups, with contemporary painting tending to reflect the interests of those groups. For their part, artists may face a complex set of personal and social considerations if they wish their art to be seen and influential, or even just to find a market for it. The artist, Jill Bray, was a resident who joined the protest through love of the local countryside. Her aim in the picture was to represent feelings generated at a particular moment at the start of the protest. A circle of chalk dust was scattered on the grass downland in the route of the proposed road and the protestors stood round in the circle joining hands in solidarity. She attempted to capture that feeling as a thing of value in itself, regardless of how many people saw the painting (Bray, personal communication). For the art-literate audience, the answer is less simple. For some, the painting could be interpreted as documenting an era or as propaganda for a political cause. For others, it would be judged and evaluated against canons of art criticism. For yet others, the painting could represent an investment opportunity if it were later to come to market.

Like the painting, the poem was not necessarily written with wide publication in mind. Nevertheless, Kate Evans chose this poem and others by Hattie to illustrate her book *Copse: the cartoon book of tree protesting* (Evans, 1998) because, in her judgement, it said something that other road protestors would agree with and wish to share. Though some of the language is colourful and contemporary, careful reading of the poem recognises long-standing poetic conventions and indicates language used with some subtlety. Indeed the poem adopted a convention of declamatory address that has a formal history in poetry dating back over 2,000 years. Its impact lay in its perceived ability to represent the feelings of a wider community of activists. Though written by one person, it employed collective nouns throughout: *we* (the protestors) are speaking to *you* (the forces of authority); *we* are a collective of frail feeling individuals, whereas *you* are a singular objectified enemy without humanity. In this way, Hattie unselfconsciously put forward a notion of group solidarity within a rallying cry that aimed to strengthen the resolve of active protestors. Significantly, she used a formal and traditional form of communication associated with mainstream society to criticise the values of that society.

Summing up, these four representations of the road protest movement clearly address different audiences in contrasting ways. As individuals, we might well have opinions about which of these media we might choose to get our message across if we found ourselves in the position of the road protestors. For current discussion, however, the key point is that these representations *serve different purposes*. To recognise this is to begin to understand something of what they mean. Each form of representation has its own theories, methodologies, techniques and history.

Representations and methodology

Besides revealing various forms that environmental representations can take, the case study of road protesting in the United Kingdom also shows the rich *types of evidence* that they supply. The four representations indicate issues and events that concern or interest people in their own way, in their own language and in their own time. From the point of view of researching into road protest, they serve to illustrate something of the variety of ideas, images and texts circulating in society. Most important in this respect is that they were unsolicited (i.e. they were not produced as a response to a researcher asking questions) and people expressed themselves in their own way (their form and content existed independently of a researcher and were not influenced by the research question).

These insights open a series of conceptual and methodological issues on which we can build. These might begin by recognising that questions relating to environmental issues like road building and the anti-roads movement in turn point to a series of larger cultural questions. These include:

- *Fascination with the car*: why are people so attached to their motorcars? What does car ownership mean in terms of social status and cultural values?
- *Belief in science and the statements of experts*: why do statements by scientists carry the aura of objective or incontrovertible truth? How does the government or its agencies use the work of scientists to give authority to their arguments? How does the road protest movement use the work of scientists to counter those arguments?
- *The ability to forget social differences and rally round a specific symbol of communal concern*: how do coherent protest groups form from constituent elements that can include anyone from militant eco-activists to local residents concerned with property prices? Is everyone mobilised behind the same cause and protesting against the same thing? How stable are the resulting alliances?

Answering these types of questions might direct us towards selecting a further set of representations that corroborate or contrast with the original representations. Equally, they might spark off new connections and lines of inquiry. Suitable candidates for analysis might include:

- advertisements that depict the car as a symbol of individual freedom and economic status;
- news items in which scientific data such as traffic forecasts and environmental impact are treated as statements of objective truth, but which further scrutiny suggests can be challenged;
- depictions of people taking part in communal living experiments in eco-friendly surroundings or participating in protest rallies on behalf of environmental causes.

Having said this, there is no suggestion that one should examine nothing but cultural representations, or study them in isolation or be restricted to one form of methodology. The choice of qualitative or quantitative methods is important here.

Qualitative and quantitative methods

Researchers often use the terms 'quantitative' or 'qualitative' to describe research methods, although in practice it is difficult to maintain a watertight distinction between them.

The term 'quantitative' generally refers to the use of methodologies that have a statistical or mathematical basis. When using quantitative methods, researchers normally look for empirical evidence in quantifiable form – such as census (demographic) population data, economic statistics and survey data. What is important is not the data *per se*, but the fact that once collected the information is analysed statistically in line with general principles of statistical analysis, such as testing for validity and representativeness.

The term 'qualitative' is more difficult to define. It covers a wide range of methods that explore aspects of human life for which quantitative techniques are considered inappropriate. Qualitative inquiries often accept far smaller sample sizes than needed to meet the criterion of being statistically representative, but do so in order to gain greater depth of understanding of particular aspects of social, cultural or mental life. Qualitative methods include:

- in-depth interviewing;
- inquiries into life history;
- participant observation – joining groups for an extended period of time to study their thinking and behaviour at first hand;
- focus groups – groups of individuals 'selected and assembled by researchers to discuss and comment on, from personal experience, the topic that is the subject of the research' (Powell and Single, 1996, 499).

Looking back at the style of analysis adopted above, we see it is broadly qualitative: that is, it generates non-numeric data from which to address the meaning of environmental representations. Further understanding of questions related to road protest might warrant recourse to other research methodologies. Some would also involve qualitative methods, such as exploring the views of road protestors through participant observation of protest groups – where the researchers join the protestors in the field as participants for a period of time and report on what they have seen. Other types of analysis might employ quantitative methods, in which the resulting data are directly expressed in numerical form. These might include:

- Examining the structure of environmental representations through content analysis, conventionally defined as techniques for the systematic and quantitative description of measurable aspects of communicated materials. This might encompass frequency counts of the mention of the topic of interest in a particular newspaper, summations of the number of column inches devoted to that topic, or listing and categorising key words used about it. The broad assumption is that more important topics feature more often and more prominently.

- Statistically based cost–benefit analysis of actual and projected costs of the road-building programme.
- Environmental impact analysis of affected sites, with particular attention to those with 'protected status' (e.g. Sites of Special Scientific Interest or Areas of Outstanding Natural Beauty).
- Social survey findings, which might involve administering formal question-naires to the general public or to a specific interest group in order to find out their views.

Each method of inquiry yields different forms of data and, equally, each has its advantages and drawbacks when relating to the initial research questions. Care is needed, however, to avoid the pitfall of appraising the merits of different methods in terms of 'bias' versus 'objectivity' or 'image' versus 'reality'. There are several reasons for this:

- Although quantitative methodologies may produce numerically exact data, the data are not necessarily any more objective or free from 'bias' than the data produced by qualitative methods. Both forms of data invariably reflect the values that framed the initial questions posed by the researcher.
- More fundamentally, *all* forms of cultural representations, regardless of the media through which they are communicated, are shaped in one way or another by our opinions and values.
- In that context, while the term 'bias' is convenient shorthand for strongly held opinions, it has no analytical value in the study of cultural repre-sentations. The designation of particular opinions as being either socially acceptable and *normal* or socially unacceptable and *biased* changes over time. What is considered socially acceptable and normal in one era may appear different over time as society changes, as, for example, with changing attitudes towards the family in the last 40 years. There is, therefore, no single absolute measure of normality against which to compare the opinions of others.
- Research into cultural representations tries to take seriously the ways in which people think about themselves and the environments in which they live. To dismiss the views of individuals or groups as 'biased', or merely myths and illusions, signifies a failure to recognise that these are the values by which specific groups organise reality or have reality organised for them.

This does not mean, of course, that all opinions are equally valid at all times or that researchers should abandon their own opinions as being distortions likely to invalidate research findings. It is perfectly possible to take other people's views seriously in and on their own terms, while remaining aware of one's own attitudes and values. Indeed, it is also valid to scrutinise the views of others and make critical comments about them.

Following that line in the case of media reporting of violent conflicts between road protestors, the police and security guards, for example, would mean providing some sense of the issues and activities that motivate those creating particular environmental representations. For instance, an anti-road protest group might exaggerate the level of violence experienced at a particular demonstration because it will generate more sympathy for their cause if the media cast the road builders and government as authoritarian bullies. For their part, government and road builders might encourage similar media reports if they thought this might create an image of the road protestors as an unruly rabble motivated by political extremism. To make these points is not to level accusations of bias, but to understand *why* each side says certain things and acts in particular ways.

Making sense of representation

So far, we have treated 'representation' as an unproblematic notion that embraces depictions of places, regions and landscapes drawn from a wide variety of media. That position now needs to be qualified. Certainly what we see or read is affected by the processes by which marks on paper, patterns of light on screen, paint on canvas, the shaping of sculptural materials or the fixing of chemicals on photographic paper create a taken-for-granted substitute for 'the real world'. As a starting point, it is worth considering the illustration shown in Figure 2.5. This drawing is taken from a booklet produced by the United Kingdom's Forestry Commission to explain its Forest Parks to the general public (see Revill and Watkins, 1996). It represented a scene in Glen More Forest Park in Scotland's Cairngorms (Edlin, 1969, 37). Drawn by a regional artist, Conrad McKenna, it depicted a forest and moorland scene rendered as a woodcut illustration. (A woodcut is a print where the image is formed by chiselling marks into a flat block of wood, which is then covered in printing ink and pressed onto a sheet of paper, where it leaves the traces of ink that make the picture.) The picture features many individual images – people, specific birds and animals, different types of tree and a background setting of lake and mountains. A capercailzie (a bird also known as the wood-grouse), two roe deer and a patch of indeterminate woodland flowers occupy the foreground. A Scots pine dominates the middle distance. A male and female hiker, dressed in tartan clothing, occupy left-centre. Their posture suggests that they are enjoying the view of the loch (lake), its wooded shores and the distant hills that complete the composition.

Figure 2.5 *Cairngorm scene: crag, loch, pinewood, capercailzie and roe deer*

Source: Black and white illustration from *National Forest Parks* (p. 37) published by the Forestry Commission, 1969. © Crown copyright material is reproduced with the permission of the Controller of HMSO and Queen's Printer for Scotland

At a superficial level, Figure 2.5 simply portrays a scene of the wildlife and vegetation awaiting the visitor, but each element of the picture is both something individual and part of the greater whole. With greater familiarity of traditional ways of depicting Highland Scotland, or notions of Scottish tourism, or even the self-perceived role of the Forestry Commission, other meanings become apparent. It then becomes clear that this illustration does not have simple and transparent meanings about which there is universal agreement. Rather, there are many different layers of meaning that can be extracted.

Exercise 2.2

Look at the bird in the bottom left-hand corner of Figure 2.5. The following are three sentences which might be used to describe this part of the picture:

- This *is a good likeness for* a capercailzie; the artist is particularly skilled at drawing birds.
- This *is just another picture of* a capercailzie, one of many in this and other guidebooks on Scotland.
- This *tells us about* the capercailzie and its threatened habitat; it is important for raising public awareness of this neglected species.

Now replace the words in italic in these sentences with the word 'represents'.

a. Can the word 'represents' be used equally in each of these sentences?

b. What other words could you use instead of the word 'represents'? If in doubt, three words that you might try are:

- shows;
- exemplifies;
- speaks for.

This exercise clearly emphasises three of the meanings that the term 'representation' was shown to have in Chapter 1:

- as a reproduction or likeness of something, such as an object, a person, place, event or idea;
- as a material or symbolic creation that exemplifies or typifies something, often designed to convey the essential characteristics of a wider set of phenomena;
- as the act of putting forward images and other pieces of evidence as part of a structured argument aimed at putting across the viewpoint of a particular individual or group – usually in the hope of bringing about change.

These characteristics often occur together. There are many instances when illustrating something becomes inextricably associated with the acts of exemplifying it and speaking on its behalf. One means to bring them together, however, is to act on Stuart Hall's advice that: 'representation is the production of meaning through language' (Hall, 1997). In saying that, Hall did not mean that only speech and written text could represent things. Rather, he recognised that all human cultural activity communicates and that all forms of communication can be thought as purposeful arrangements of signs and symbols. These can be decoded in the same way that we might try to understand an unfamiliar language.

Students of cultural representations, for example, may examine clothing, gesture and facial expressions using the same conceptual framework that deals with written language or visual images. They would argue that clothing and gesture are media by which humans communicate with each other, for example, expressing affection, anger, pride or pleasure. By doing so, they constitute conscious arrangements of signs and symbols that can be analysed as forms of language.

Moreover, whichever type of representation is under scrutiny, it is valuable to think of two elements of language that are centrally involved in the production of meaning. The first is *structure*, which concerns the form and content of representations. The second is *context*, which concerns the setting in which the representation is located. While researchers argue about how far it is possible to differentiate between structure and context when reading and interpreting representations, it is a useful distinction to make when initially exploring the processes by which representations represent. For the rest of this chapter, therefore, we concentrate on and exemplify issues concerning structure and structural processes. In Chapter 3, we turn to context and contextual processes.

Signs and symbols

The study of semiotics provides a valuable point of entry when examining structural components of representation, which consist of the individual elements within a representation and the relationships between them.

Semiotics and semiology

The words 'semiotics' and 'semiology' first originated in medical parlance in the seventeenth century as interchangeable terms connected with the science of interpreting symptoms. Their use in connection with linguistic science, where they are again interchangeable, developed in the early twentieth century, particularly through the work of the Swiss linguistic theorist Ferdinand de Saussure (1857–1913) and the French cultural theorist Roland Barthes (1915–80). 'Semiotics', the word used in this text, is defined as the 'science of signification' (Watt, 1998, 675) or, rather more fully, 'the science of communication studied through the interpretation of signs and symbols as they operate in various fields, especially language' (*Oxford English Dictionary*, xiv, 958). Its central aim is to show how one thing can mean another. In other words, how on seeing 'x' can someone be induced to think of 'y' even though 'y' is absent (Watt, 1998, 675)?

The answer to this question lies essentially in understanding the process by which things in the world relate to ideas in our minds. From a semiotic perspective, there are two processes involved: assigning symbols to things and assigning symbols to ideas. According to de Saussure, *signs* (symbols which stand for or take the place of something in the outside world or the mind) are formed from two elements: the form (the actual word or image) and the corresponding idea or mental concept with which the form was associated. The first element is the *signifier*; the second is the *signified*. Each time someone hears, reads or sees the signifier, it correlates with the signified. Both are necessary to produce meaning but representation rests on the relation between them, fixed by our cultural and linguistic codes. In this way: 'though we may speak . . . as if they are separate entities, they exist only as components of the sign . . . (which is) the central fact of language' (Culler, 1976; quoted in Hall, 1997, 31). This relationship is often described using the following simple equation:

$$\text{Signifier} + \text{Signified} = \text{Sign}$$

Signs themselves may be classified. Although a number of different classification systems exist, the most widely used is perhaps that devised by Barthes (e.g. 1967, 1972; see Culler, 1983), who divided signification – the process by which meaning is given to signs – into two forms. The first, denotation, is a form of symbolism in which there is a direct relationship between the sign and the thing it represents (similar to 'metonymy', see below, p. 33). The second connotation is a form of symbolism in which there is an indirect relationship between the sign and the thing it represents (similar to 'metaphor', see below, p. 33). Denotative signs, for instance, would include a picture of a tree, which resembles or reproduces in some way an actual tree in the landscape. By contrast, connotative signs might include a picture of a tree to represent vitality, growth and life.

This analysis offers ideas that help when examining the sort of messages contained in texts or visual images and when considering the way that the message is communicated to the audience. It supplies:

1   A *conceptual framework* using the notions of signifier, signified and sign.
2   A *hierarchy of meaning*. Barthes' distinction between denotation and connotation suggests that there are at least two levels of signification. These are:

   ● that which is denoted (signifier): what the representation actually (literally) shows;

- that which is connoted (signified): what the image causes the audience to think about, what it implies (metaphorically) when the representation is seen or read.

Some idea of how to apply these ideas can be gained from further analysis of Figure 2.5.

Exercise 2.3

Make a table with three columns headed, respectively, 'signifier', 'signified' and 'sign'. Now look again at the scene depicted in Figure 2.5. Alongside these make a list of the thing (signifier), idea or concept (signified) with which you think these may be connected.

To give you an example:

    signifier – the man pointing
    signified – tourist or rambler
    sign – adventure/discovery

If you find this difficult, remember that you already have some clues as to the kinds of meaning that might be being suggested here. We know this is a picture in a guidebook commissioned by the United Kingdom's Forestry Commission to promote tourism in one of their Forest Parks in Scotland. Ask yourself: 'If I represented the Forestry Commission, what ideas might I want to get across?'

Table 2.1 *Cairngorm scene: signifiers, signifieds and signs*

Signifier	Signified	Sign
Scots pine	Notable Scottish species	Highland ecology
Kilts, Highland clothing	Characteristic Scottish clothing	Scottish tradition
Rock promontory	Good vantage points	Opportunity to encounter spectacular scenery
Loch and distant hills	Typical Highland scenery	Tranquillity, grandeur
Capercailzie	Wildlife	Opportunity to encounter nature

The list supplied in Table 2.1 is not exhaustive and the interpretations given are not incontrovertible, but it shows how the semiotic framework can help to identify different levels of meaning in this apparently straightforward drawing.

Four other notions, that recur continually when examining signs as cultural representations, help to take this discussion further. They are:

1   *Anchored meaning*: where a title or written description is used to fix the meaning of a sign, text or image that has an ambiguous meaning or is clearly open to a multiplicity of interpretations. This is a way in which producers of cultural representations try to guide the audience towards accepting the meaning preferred by the maker of the representation. Figure 2.5, for example, bears the caption 'Cairngorm scene: crag, loch, pinewood, capercailzie and roe deer'. This seemingly anodyne title imparts considerable information about the picture, its purpose and how the Forestry Commission then wished readers to interpret the image. It told readers that this was a picture of a particular place in Scotland and invited them to direct their attention to the natural scenery, flora and fauna.

2   *Stereotypes*: or images that accentuate or personify particular taken-for-granted characteristics of a people or place. Sometimes they are deliberate caricatures which emphasise particular characteristics for humorous effect. Sometimes they are not premeditated, but result from the unreflected prejudices and preconceptions of the image-maker. Researchers sometimes find the latter the most interesting form of stereotype because they hint at values and assumptions of which the creators of the representation may not even be aware. The 'commonsense' values that inform stereotypes are the values about which social scientists are most critical because they reflect the taken-for-granted values that create inequalities in society, notably including those concerning issues of race, class and gender.

Exercise 2.4

Look at the portrayal of the male and female figures in Figure 2.5. How might this be regarded as stereotyped?

The drawing shows the male leading the female figure, who gazes passively as he points into the distance. This portrayal of males as dominant, knowledgeable and adventurous and females as passive, timid and dependent reflects what were once widely held assumptions about the respective roles of men and women. It is likely that Conrad McKenna was unaware of the gender stereotyping that he drew into this representation and that therefore this reflects his unreflected prejudices and preconceptions. Certainly, there is no technical reason why the drawing could not show the roles reversed, with the woman depicted as the dominant figure.

3   *Metaphor*: or a symbol that stands for something else. With metaphorical images, there is a kind of poetic relationship between the image and the thing that it symbolises. Figure 2.5 contains many examples, such as the tartan as a metaphor for Scottishness, the flowers in the foreground as a metaphor for fragile, delicate nature, and the capercailzie and roe deer (both animals kept as game) as a metaphor for hunting.

4   *Metonymy*: like metaphor, metonymic images are symbols of something else but, in this case, a part of something stands as a symbol that represents the whole. Examples of metonymy abound in Figure 2.5, such as the use of a single Scots pine in the foreground as a metonymy for native woodland or the portrayal of a mountain in the background as an exemplar of the Cairngorms, the range of rugged mountains in which this Forest Park is set. There is often no neat distinction between metonymy and metaphor. As an indigenous bird – even if one driven to extinction and then reintroduced from Scandinavia for hunting purposes – the capercailzie is a metaphor for Scottishness, but also a metonymy for a distinctive Highland ecosystem.

This question of overlap has conceptual as well as technical implications for our analysis. With greater familiarity, it becomes progressively harder to isolate the meaning of one part of the picture from the others, since its meaning increasingly makes sense only in terms of other elements in that picture. When first looking at Figure 2.5, we noted that this picture comprises many discrete elements yet the whole is greater than the sum of its component parts. The various elements work together to tell a bigger story. Semioticians describe this as its *rhetorical structure*, using the term 'rhetorical' to refer to the ability of a particular representation to get a specific message across.

Exercise 2.5

Returning again to Figure 2.5, make a list of individual picture elements that work together to reinforce a common message.

If you find this task difficult, use the following three concepts as an organising structure:

    Scottishness
    wilderness
    tranquillity

Given what we already know about why the picture shown in Figure 2.5 was produced and how it was used, it is now possible to piece together a rhetorical structure and an overall message for this picture. Using the three headings given above, we can start to see how the image works. To the left-hand side and towards the foreground are found a range of signs representing Scottishness – the caper-cailzie, the Scots pine and the two people dressed in tartan. In the distance are seen signs representing a rugged and wild landscape, mountains and valleys, billowing clouds and extensive forest areas. To the right of the picture in the middle distance and foreground are signs representing peace, tranquillity and the delicate, aspects of nature – flowers, deer and a lake of calm water. Viewed from our elevated perspective, the readers (as potential tourists or hikers) are represented in the picture by the two figures, one of whom points out the scene to them as much as to his companion. The illustration invited readers to place themselves in the shoes of these hikers and to enjoy the countryside in a similar manner. Thus the reader was invited to participate in a range of rural environmental experiences (adventure, relaxation, discovery, pleasure) that, the Forestry Commission claimed, are all available in this particular Forest Park.

Representations and audiences

To this point, we have focused on meanings suggested by the intentions of those who produce representations, but have also shown that active participation is needed if a representation is to have meaning for its audience. From the earlier road protest case studies, reader involvement is most apparent in the poem (Figure 2.4) or the painting (Figure 2.2) where the meaning seems to derive from the con-nections made with the audience's thoughts and experience. Yet the same process applies even for representations that appear to have little other than literal meaning. The newspaper photograph (Figure 2.3), for example, might be interpreted as a comment on the destruction caused by road building, but if read literally is only an arrangement of images of road, traffic, hillside and excavation work. In isolation, these impart very little; it is only in the context of the audience's wider experience that the pieces are pulled together. The reader might have seen an item about the road protest story on television, discussed it with friends, have opinions about the road-building programme in general, or have driven along this stretch of road and have thoughts about the need for this specific development. No matter how those who create interpretations try to enforce a particular reading, they cannot control the circumstances in which the audience receives and consumes the message. Those who produce representations can never prevent audiences reading their own thoughts and experiences into the representation. Meaning is at least as much a matter of audience interpretation as the producer's intention.

Exercise 2.6

Returning finally to Figure 2.5, place yourself in the position of a potential tourist to Scotland. Would the picture shown be effective in encouraging you to go walking in the Cairngorms? Give reasons for your decision.

Now read the following critical comments that were made in a general assessment of the Forest Park guidebooks published in 1968, when these guidebooks were still in active use:

> There is an obvious risk that the interests of the forest manager, which commonly may lean towards natural history, tend to be magnified in the assumed demands and interests of the public users. This is readily apparent in the form and content of the guidebook publications of the Forestry Commission. The guides are lengthy and descriptive, botanical, zoological and archaeological in matter, and are intended for use in conjunction with a 1-inch or larger scale map. Undoubtedly they are absorbingly interesting guidebooks for the fairly small proportion of walking-holiday visitors, but they make an insignificant impact on the car-driving, day-visiting family parties who form the majority of the present visitors.
>
> (Mutch, 1968, 83)

To what extent do these criticisms of the Forest Park guides agree with your opinion of the effectiveness of the picture as an image intended to encourage tourists?

Exercise 2.6 shows something of the complexity involved in trying to decipher the meaning that representations have to their audiences. Your answers to this exercise rely on judgements made at the present day, whereas this picture comes from a guidebook first printed in the 1950s and withdrawn from circulation in the early 1970s. Understandably, it is now difficult to ask people whether the picture was effective or not. It relates to a time many years ago and, even if it were possible to find people who used this guidebook, there is no guarantee that their recollections would accurately reflect how they felt at the time. The comments reproduced above also relate to the series of guidebooks as a whole, not to the specific picture shown in Figure 2.5. Moreover, although the author notes that his conclusions are based on independently commissioned survey work, it is unclear to what extent these particular conclusions were based on any specific research, such as survey questions, or were just the writer's independent judgements. We certainly need to know more about how these conclusions were reached before accepting the author's conclusions at face value. The conceptual and methodological issues involved are studied in the next chapter, where we examine issues raised by the wider social context of environmental representations.

Further reading

On the basis of representation as a cultural practice, see:

Stuart Hall, ed. (1997) *Representation: cultural representations and signifying practices*, London: Sage/Open University.

For a critical introduction to a range of techniques of visual analysis, see:

Gillian Rose (2001) *Visual Methodologies*, London: Sage.
Theo Van Leeuwen and Carey Jewitt, eds (2001) *Handbook of Visual Analysis*, London: Sage.

3 Representations in context

This chapter:

- examines strategies for studying the audience's consumption of environmental representations;
- considers the contexts in which representations are produced;
- explores the concepts of cultural politics, ideology and discourse;
- provides a range of basic approaches for the study of the discursive role of representations in society.

Audiences and the media

The media occupy a central place in any study of environmental representations. Traditionally, 'media' meant the 'media of mass communication', by which technologically based systems transmitted content (or messages) through print, broadcasting, posters or film to remote and scattered audiences. More recently, though, the idea of what constitutes 'media' has expanded dramatically. If media transmit content in symbolic form to an audience, then architecture, clothes, recorded music, food packaging, jewellery and skin tattoos are legitimately media. Moreover, many new media no longer conform to the established model. The convergence of television, computers, wireless technology, digital networks and mobile telephones brings patterns of use that bear little resemblance to the 'fireside' consumption of early radio and television programmes.

The word 'consumption' needs clarification. We stressed the importance of the audience's active participation in shaping the meaning of environmental representations at the end of Chapter 2. In this context, it might seem that media research has much to offer, given that audience studies have preoccupied media researchers since the 1920s (McQuail, 1997). Sadly, however, the findings of that research have proven continually disappointing. Researchers have often attempted to trace direct and measurable audience response to communication, usually finishing by either showing no discernible effects or recognising that any effects are very difficult to assess and quantify. The problem lies partly with conceptual under-standing and partly with methodology. In terms of the former, researchers reluctantly accept that the media do not function in isolation and that, in most circumstances, any impact that they might have is itself *mediated* by many other social and cultural factors. In terms of the latter, some of the problems certainly lay in the methods used to study the relationship. Intuitively one might think that the best way to find out what an audience thinks of a particular representation is to go out and ask them, but the process is fraught with problems. These include:

1   *Researcher effects*. At the very least, questionnaire and interview schedules direct the responses of interviewees towards issues about which they may not be particularly interested or have thought little about previously. Undue emphasis may be given to issues that are not necessarily important to the everyday lives of respondents. Though there are techniques for minimising the problem, the yes/no, ranked and multiple response question strategies used in questionnaires constrain individual choices of answer. Critics argue, for example, that questionnaires produce the answers that interviewers expect rather than the ones respondents might give if not guided. Indeed in the case of relatively private leisure pursuits such as watching television or listening to radio, the very presence or interest of a researcher may change the attitudes and behaviours of listeners and viewers. The likelihood is that this will encourage them to try to reflect researcher expectations and, in the process, conform to perceived cultural norms.

2   *Causality and the media environment*. Much audience research has con-centrated on trying to appraise the specific effects of a particular media event or group of events on the public. Examples might include assessing the effectiveness of a particular advertising campaign or trying to discover if exposure to violent images on television makes viewers more aggressive. Yet, as already noted, direct causal relationships traceable to specific media events are difficult to find. We might ask then why researchers persevere with this style of research. The reason often lies in the way that the research is commissioned. Much professional media research stems from commissions by advertisers, government, the military, and market and electoral research agencies that often want to know about the 'effectiveness' of a particular isolated campaign rather than ask deeper questions. While that imperative remains, it is likely that the same flawed research design will persist.

3   *Representativeness.* Social science surveys seek to generalise from sample populations about issues that may not loom large in the popular imagination and may well be couched in language unfamiliar to those asked to respond. As a result, questions relating to representativeness are central to many of these forms of study. Much audience research is commissioned by large commercial and media organisations that have sufficiently large budgets to employ teams of researchers to produce statistically representative samples. However, representativeness is not just a matter of resources, since it also involves questions relating to the timing and location of research. If research seeks to reproduce the effects of respondents' exposure to a particular cultural representation, then they should do so in settings that replicate the normal experience of their subjects when exposed to that particular medium. This might include, for example, reading the paper on the train, listening to the radio over breakfast, or looking at a picture in a gallery. Even if researchers can obtain responses at these times, their presence is likely to be invasive and inappropriate. It is likely, therefore, that survey results seldom fully represent experience of environmental representations and their meanings as they are routinely encountered in daily life.

These considerations apply when assessing the value of the three of the major forms of survey methods used: formal structured questionnaires; viewing or listening logs; and informal and semi-structured interviews:

● *Formal structured questionnaires*: extensively used by market research companies and others on behalf of advertisers, television and radio companies to assess audience share, listening and viewing patterns, and product awareness. As noted above, researchers typically administer these questionnaires to a large sample, using randomised procedures to obtain a representative sample across the general population. Alternatively, they can target questionnaires at a particular audience group – sometimes using a regular panel of respondents in order to obtain results over a period of time (longitudinal data).
● *Viewing or listening logs*: involving respondents recording their viewing, listening or reading patterns over the course of a period of time, often a week, in the form of a diary.
● *Informal and semi-structured interviews*: including one-to-one interviews that explore in detail the attitudes and behavioural patterns of a particular small-scale, but carefully targeted sample. Focus groups (see Chapter 2) are a variant of this practice, being extensively used by media, political and campaigning organisations to test public opinion.

Formal questionnaires still have a place in audience research, but the growth of logs and informal and semi-structured interview methods reflects both the shortcomings of audience reception studies and the growing sophistication of media methods. For example, recent work in cultural studies suggests that reading media messages is a richer and more active process than previously assumed, with researchers now preferring to study audience *consumption* rather than reception.

With the movement towards recognising greater complexity rather than forever searching for often non-existent linear flows has come greater scrutiny of the questions that researchers ask. When constructing an inquiry into a specific topic, for example, one cannot ask people to dismiss everything they have ever said, read, heard or seen in order to focus on the one item or environmental representation that interests the researcher. Even if respondents are willing to attempt this somewhat unlikely task, it is impossible to ensure that they are not subconsciously influenced by representations, thoughts and ideas derived from other events, times and places. Certainly, it is extremely difficult even to identify, let alone measure, the specific changes in attitudes and behaviour linked to any single item of communication.

Having said this, audience interpretations remain important in any study of environmental representations. While not an alternative to an understanding of how representations are structured and work rhetorically, they assist in appreciating the social context of specific representations. Indeed they are particularly useful as part of approaches that draw on the technique of *triangulation*, which apply several research methodologies in combination when studying the same phenomenon. Triangulation can be employed in either quantitative or qualitative studies and, as with the use of triangulation in surveying, the most reliable results come when separate sets of findings converge. This helps to overcome the weaknesses and the problems that come from single method, single observer, or single theory studies. The purpose of triangulation in specific contexts, then, is to confirm findings through convergence of different perspectives, and there are many forms of this technique (Bryman, 2001; Hannibal, 2001). Here we compare the meanings derived from studying the overt content of representations with the meanings derived from them with other sources that may corroborate or contradict them. The aim is not to contrast cultural images with social realities, but to recognise the significance of the social context within which meanings are *negotiated* (i.e. evolve through a process whereby certain aspects of meanings are accepted and others rejected by the various actors and agencies present within society).

Signs, codes and signification

If meaning are actively 'read' or 'interpreted', even for those denotative signs regarded as literal representations of things in the outside world (see Chapter 2), then the question of meaning is inevitably more complex than often admitted.

Many argue that all signs are more or less connotative, in the sense of raising ideas and associations that convey meaning, and create their meaning through indirect associations rather than literal resemblance. This certainly seems true for spoken and written language, apparently the most 'natural' form of communication. Here a system of hieroglyphs or symbols represents collections of sounds strung together to form words that, in turn, are assembled to form phrases, sentences and paragraphs. There is rarely a direct imitative relationship in language between symbols and their meanings. Only in special cases called onomatopoeia do words resemble the things they signify ('woof' for English speakers signifies a dog bark; 'ring' signifies the sound of a bell). One has only to look at the very different arrangement of letters and sounds used to represent simple commonplace things like 'dog' or 'tree' in geographically adjacent languages such as English, French and German to recognise that, by and large, the symbols with which people communicate are *arbitrary*. In other words, there is no necessary or inevitable relationship between signs and the things that they signify. Furthermore, the meanings of representations are constantly subject to interpretation and reinterpretation. Stuart Hall (1997, 32) noted the far-reaching implications of this argument for theorising representation and understanding culture:

> Words shift their meanings. The concepts (signifieds) to which they refer also change, historically, and every shift alters the conceptual map of the culture, leading different cultures, at different historical moments, to classify and think about the world differently.

Exercise 3.1

Consider the following two representations of the impact of industrialisation in the environment during the Industrial Revolution in Britain.

The first, shown in Figure 3.1, is a watercolour painting entitled *Dudley, Worcestershire* by J.M.W. Turner (1775–1851), who, along with his contemporary John Constable (1776–1837), is probably the most famous of English landscape painters. Dudley had been an important metalworking centre since the medieval period and at the time when Turner visited in 1830 the iron industry had diversified to include casting and finishing, as well as chain-making (Rodner, 1997, 108). Engraved in 1835 by Robert Wallis, the picture appeared three years later in the collected two-volume edition of 96 of the artist's prints, *Picturesque Views in England and Wales*. This picture shows the canal with loaded narrow boats in the foreground, pottery kilns, iron works and coal mines in the middle distance and Dudley, the town with its church overlooked by Dudley Castle in the background.

The second is the famous depiction of the woollen textile manufacturing villages in the area adjacent to the Yorkshire town of Halifax by Daniel Defoe (1660–1731). It

continued

was written during the early years of industrial expansion and technological change in England. The passage is taken from his *A Tour through the Whole Island of Great Britain*, published in three volumes during the period 1724–7. The book is substantially based on his own travels through England and in Scotland that began in the early 1700s.

Make two short lists itemising evidence for environmental devastation brought on by the unchecked growth of industrialisation in these two sources.

Do you think Turner and Defoe see industrialisation as having positive or negative effects on the environment?

Figure 3.1 Dudley, Worcestershire *(1835)* by J.M.W. Turner *(1775–1851)*. *Courtesy of the Board of Trustees of the National Museums and Galleries on Merseyside (Lady Lever Art Gallery, Port Sunlight)*

But now I must observe to you, that after having pass'd the second hill, and come down into the valley again, and so still the nearer we came to Halifax, we found the houses thicker, and the villages greater in every bottom; and not only so, but the sides of the hills, which were very steep every way, were spread with houses, and that very thick; for the land being divided into small enclosures, that is to say, from two acres to six or seven acres each, seldom more; every three or four pieces of land had a house belonging to it.

Then it was I began to perceive the reason and nature of the thing, and found that this division of the land into small pieces, and scattering of the dwellings, was occasioned by, and done for the convenience of the business which the people were generally employ'd in, and that, as I said before, though we saw no people; those people all full of business; not a beggar, not an idle person to be seen, except here and there an alms-house, where people antient, decreptid, and past labour, might perhaps be found; for it is observable, that the people here, however, labourious, generally live to a great age, a certain testimony to the goodness and wholesomeness of the country, which is without doubt, as healthy as any part of England; nor is the health of the people lessn'd, but help'd and established by their being constantly employ'd, and, as we call it, their working hard; so that they find a double advantage by their being always in business.

This business is the clothing trade, for the convenience of which the houses are thus scattered and spread upon the side of the hills, as above, even from the bottom to the top; the reason is this; such has been the bounty of nature in this otherwise frightful country, that two things essential to the business, as well as to the ease of the people are found here, and that in a situation which I never saw the like of in any part of England; and, I believe, the like is not to be seen so contrived in any part of the world; I mean coals and running water upon the tops of the highest hills: This seems to have been directed by the wise hand of Providence for the very purpose which is now served by it, namely, the manufacturers, which otherwise could not be carried on; neither indeed could one fifth part of the inhabitants be supported without them. After we had mounted the third hill, we found the country, in short, one continued village, tho' mountainous every way, as before; hardly a house standing out of a speaking distance from another, and (which soon told us their business) the day clearing up, and the sun glancing, and as I may say, shining (the white reflecting its rays) to us, I thought it was the most agreeable sight that I ever saw, for the hills, as I say, rising and falling so thick, and the vallies opening sometimes like streets near St. Giles's called the Seven Dials; we could see through the glades almost every way round us, yet look which way we would, high to the tops, and low to the bottoms, it was all the same; innumerable houses and tenters, and a white piece upon every tenter [tenters are frames on which cloth is stretched and dried].

But to return to the reason of dispersing the houses, as above; I found as our road pass'd among them, for indeed no road could do otherwise, wherever we pass'd any house we found a little rill or gutter of running water, if the house was above the road, it cam from it, and cross'd the way to run to another; if the house was below us, it cross'd us from some other distant house above it, and at every considerable house was a manufactory or work-house, and as they could not do their business without water, the little streams were so parted and guided by gutters or pipes, and by pipes, and by turning and dividing the streams, that none of those houses were without a river, if I may call it so, running into and through their work-houses.

Again, as the dying-houses, scouring-shops and places where they used this water, emitted the water again, ting'd with the drugs of the dying fat, and with the oil, the soap, the tallow, and other ingredients used by the clothiers in dressing and scouring, &c. which then runs away thro' the lands to the next, the grounds are not only universally waterd, how dry soever the season, but that water so ting'd and so fatten'd enriches the lands they run through, that 'tis hardly to be imagined how fertile and rich the soil is made by it.

(Daniel Defoe, 1724–7)

Having completed this exercise, your lists may include some of the following:

Turner's Dudley

- Smoke and atmospheric pollution form a dark haze over the town obscuring the view of the moon and blotting out the historic church and castle.
- Visual invasion of unregulated ugly industrial development into historic built environment of the town.
- Industries working round the clock suggesting long working hours for workers and relentless intrusion of noise, smoke and light pollution day and night.

Defoe's Halifax

- Houses increasingly close together (forming a conurbation).
- Fields sub-divided and turned over to industrial use (drying cloth).
- Streams and watercourses diverted and sub-divided to serve industry.
- Unchecked pollution of water courses with oil, soap, tallow and other industrial effluents.

It would be easy to say that both Turner and Defoe saw the environmental impacts of industrialisation negatively, but to do so without qualification imposes a twenty-first-century value system on representations that are, respectively, well over 250 and 150 years old. The conventions routinely adopted to interpret these representations today are not necessarily appropriate. In order to understand more about these representations and what they mean, it is necessary to set them in the context of what is known about the environmental values and relations of the time as well as finding out more about the writer and the artist concerned. It is not feasible to undertake an audience reception questionnaire even if that were desirable, but it is certainly possible to gather evidence that supplies a geographical and historical context in which to interpret these environmental representations. Among the questions that need to be asked are:

1   The market: for whom was the representation produced? Who bought the representation?
2   The audience: how did audiences and critics receive it when first shown?
3   How have interpretations of the representation changed over time, or from place to place?
4   The artists or writers: what characterised their output in terms of style and subject matter? What were their views of their own work, its form and purpose? How does this work fit with their other work in terms of representativeness, continuity and change?

Sources for this sort of information can be included under two headings; on the one hand, critical interpretations, and on the other, indirect measures and descriptive statistics.

Critical interpretations include:

●   Writings in books, journals or newspapers about the artist, writer, book or painting contemporary with its production.
●   Writings about the artist, writer, book or painting produced in subsequent periods.
●   Reviews of exhibitions.
●   Present-day academic research.
●   Writings, personal manifestos, autobiographical testimony, diaries and lectures by the artist or writer.

Indirect measures and descriptive statistics include:

●   Numbers of editions and copies of the work sold at the time and subsequently.
●   Popularity of gallery or exhibitions in which the picture was displayed.
●   Audience or market sector at which reproductions of the book or painting are targeted.
●   Patronage: did an individual or an organisation commission the representation? If so, do we know why they commissioned it? Did the artist or writer produce it speculatively and offer it for sale? Did the artist or writer produce it for his or her own purposes and refuse to sell it?
●   Evidence from the artist's or writer's other work. Is this particular representation typical, exceptional or indicative or a change in their work in terms of style, structure and subject matter?

It would be very fortunate to obtain answers to all these questions, but answers at least to some are necessary for the researcher to arrive at an informed interpretation of the representation.

Exercise 3.2

Do the following statements change the way we might interpret Defoe's writings or Turner's painting?

Defoe:

- believed that all people including children should work or be otherwise gainfully occupied;
- did not like barren landscapes (highlands or moorlands) that could not support human habitation or be economically exploited;
- explicitly stated in the preface that he wrote his *Tour* specifically on the subject of Britain's wealth, prosperity and 'increase';
- was employed as a political spy, propagandist and writer latterly in the service of the 'Whigs', a political grouping concerned with expanding trade, industry, science, learning, individual self-reliance and paternal welfare as sources of patriotic strength (see also Chapter 5).

Turner:

- painted pictures of railways and rivers which emphasised the positive role of modern communications;
- interested himself in new developments in science and technology that also had an impact on his painting;
- would have known that the owners of Dudley Castle were also important industrial entrepreneurs and played a fundamental role in the development of industry in and around Dudley.
- worked at a time when the new middle class of professionals and business people found the power of new industrial processes both shockingly new and awe inspiringly majestic.

Re-examining the reading by Defoe in light of this additional information, it is apparent that although Defoe itemised what may now seem negative aspects of industrialisation, this was not his main focus. He wanted to tell his readers how the transformations wrought by industrialisation brought many benefits for the population. For Defoe and his political allies, this 'state of the nation' type report was as much a matter of political argument as dispassionate documentary. Defoe saw industry as 'the bounty of nature in this otherwise frightful country'. We can understand this when told that he disliked barren landscapes such as highlands and moorlands, which did not support human habitation. Here 'the wise hand of providence' gave mineral wealth and the population had a duty to exploit those resources and create a busy civilisation from an unproductive, and therefore visually unattractive, hilly countryside. There was a clear link between the perceived visual attractiveness of the environment and Defoe's advocacy of particular social and economic values. Perhaps surprisingly for present-day thinking, Defoe thought that hillsides thickly covered with industrial activity were 'the most agreeable sight that I ever saw' shining white in the sunshine. The new urban landscape encroaching ever further into the dales and moorlands of the Pennine hills reminded him favourably of city streets in London. Quite astonishingly, he even used his visual aesthetic of industry and business to justify evidence of water pollution:

> that water so ting'd and so fatten'd enriches the lands they run through, that 'tis hardly to be imagined how fertile and rich the soil is made by it.

Defoe, therefore, seemed to think that as human industry, itself a social good and bringer of material prosperity, produced these waste products, they in turn must have an intrinsic goodness that they imparted on the surrounding fields through which the streams flowed. Defoe's logic employed metonymy (see Chapter 2), in which a physical part of the industrial process assumed the beneficial physical qualities of industrial activity and actively imparted material prosperity on to anything with which it came into contact.

Similarly, the additional information given above informs interpretations of Dudley. Turner painted this picture towards the end of his career at a time when he experimented with studies showing processes in nature, including geology and volcanoes, storms and other atmospheric effects. The focus on atmosphere and industrial pyrotechnics fitted Turner's broader interests in the direct experience of the forces of nature. The billowing clouds of smoke and flames related to his interest in the elemental forces of the physical world rather than suggesting horror at the detrimental effects of atmospheric pollution. A further aspect of his work during this period was an emphasis on pictures that recounted stories of historical destiny.

Views of new technology supplied some clues. In 1839, Turner painted *The Fighting 'Temeraire' Tugged to her Last Berth to be Broken up, 1838*, which portrayed a steam tug towing a famous old sailing warship up the Thames to the breakers' yard. Some critics, at the time and subsequently, have interpreted this picture as showing the power of the steam age unceremoniously sweeping away traditional society, as represented by the venerable old 'man-of-war'. If this is the case, then the picture of Dudley may similarly reveal new industries and new forms of society replacing the old traditional order represented by the church and the castle. In support of this interpretation, there is much evidence to suggest that other images of industry and key sites of the Industrial Revolution were popular with the increasing numbers of the middle-class professional and commercial classes. They, like Turner, were interested in the spectacular physical energy displayed by new industries and also saw them as evidence of the importance of their own class for the nation. Yet Turner sometimes used the combination of the traditional and modern in townscapes to convey more complex relationships between old and new. Writing about Leeds, the subject of another of his paintings of an industrial town, he expressed this in poetry:

> The extended town far stretched East and West
> The high raised smoke no prototype of Rest
> Thy dim seen spires rais'd to Religion fair
> Seen first as moments thro the World of Care
> Whose Vice and Virtue so commixing blends
> Tho one returns while one destruction sends
> (quoted in Daniels, 1993, 123)

Given the context of the book of views that included the Dudley representation, it is likely that Turner knew that the Earls of Dudley, the family that owned the castle and surrounding lands, were major players in local industrial and technological development. Hence, the castle is an ambiguous symbol of tradition. Taking this into account, along with Turner's interest in technology as an expression of natural forces and his concern for continuity and tradition in society, he may well have intended to suggest a more symbiotic relationship between the authorities of social tradition and physical nature. It is possible then to interpret this painting as suggesting that nature and society are together forces for environmental change and that industry, aristocratic wealth and traditional authority are both co-dependent and a threat to one another.

From this analysis, these environmental representations not only have interesting and complex meanings in their own right, but also convey important things about how contemporaries thought about and used their environments. Moreover, each of these representations have interesting histories in which interpretations and meanings have changed in focus according to the uses to which they were put. Present-day historians have widely adopted Defoe's *Tour* as a disinterested documentary account of early eighteenth-century Britain, but this somewhat misreads his work. Defoe, the author of *Robinson Crusoe* (Defoe, 1719) among other books, was a novelist of considerable imagination. Rogers (1971, 9–10), for example, called him 'the Great Fabricator' and considered the *Tour* 'a deeply imaginative book'. Similarly Turner's picture of Dudley is widely used as a symbolic representation of the disquieting consequences of technological innovation. John Ruskin, the Victorian writer and a champion of Turner, who owned the watercolour of Dudley for a time, supported this viewpoint. Ruskin fervently criticised industrialism, industrial society and culture. In 1878 he wrote that he found in the painting a clear expression of 'what England was to become' with its 'ruined castle on the hill, and the church spire scarcely discernible among the moon-lighted clouds, as emblems of the passing away of the baron and the monk' (Rodner, 1997, 106). These are sentiments that we might understand today, but they are not views with which Turner or the contemporary middle-class audience for *Picturesque Views of England and Wales* would necessarily have sympathised.

So saying, there is no suggestion that the horrors of pollution and environmental damage created during the Industrial Revolution were mere figments of the imagi-nation. Certainly, these representations do offer observational evidence of pollution and environmental damage in eighteenth-century Halifax and nineteenth-century Dudley. Historical geographers and environmental historians routinely use representations found in pictures, diaries, eyewitness accounts, maps and surveys to help them reconstruct past environments and climatic events such as floods and droughts (see Hooke and Kain, 1982). Nevertheless, environmental representations have distinctive qualities that give them an importance that greatly transcends their role as simple documentary evidence. In particular, their content and meaning provides often remarkable insights into the values of the society in which they were produced.

Representation, meaning and values

The question of how representations and their meanings are socially produced is complex and involves issues central to our understanding of society and how it functions. At the extremes, two radically differing standpoints provide perspective on the relationship between representation, meaning and social values:

1   *Voluntarism* is the name given to a theory that considers that human will is the key to decisions and behaviour. Voluntarism supports the argument that the creation and interpretation of representations is a product of each individual. People act in society as free individuals and make their decisions based solely on their own feelings and preferences. Meaning, therefore, is a purely personal matter and individual psychology holds the key to under-standing environmental representations and their meanings. Due to the personal and private nature of interpretation, it is not possible to make connections between individual interpretations and wider social values.

2   *Determinism* is the philosophical doctrine that human action is not freely chosen but determined by external forces. Determinism supports the argument that the creation and interpretation of representations is dependent on identifiable processes and structures external to the individual. Here individuals' actions, thoughts and feelings are products of events beyond their control and often their understanding. From this perspective, the making and interpretation of representations by individuals reflects first and foremost these external forces. For researchers, analysing representations involves observing these wider social forces in operation and effect.

Exercise 3.3

Suggest voluntarist and determinist interpretations to the passage by Defoe and the scene depicted in Figure 3.1.

We would suggest that:

- Voluntarist interpretations would focus on the distinctive personal characteristics of the artist and might use these to emphasise the artist's particular form of genius. A voluntarist interpretation of Defoe's depiction of Halifax might focus then on him as both inveterate traveller and keen observer of humanity. A similar interpretation of Turner's *Dudley* could focus on his unique ability to see beauty in the diverse forces of nature.
- Determinist interpretations would ignore the personal characteristics of the artist and interpret the representations solely as the product of processes operating on society generally. Thus, Turner's *Dudley* could be understood as the product of changing economic relations that resulted in the rise to power of the new middle class of professionals and business people as patrons and purchasers of art. Similarly, Defoe's Halifax might be seen to indicate our innate instincts for the satisfaction of basic needs, which itself is reflected in an ethic of hard work.

Both these forms of interpretation are ultimately unsatisfactory. At best they explain partially; at worst they attribute universal characteristics to society and individuals (for example, in the form of instinctive behaviour) that are hard to justify for historically and geographically distant peoples. It was argued above that representations must be set in their specific context; neither voluntarist nor determinist perspectives do this adequately. Many cultural researchers have emphasised the importance of studying the relationship between human *action*, our capacity to act as free independent individuals, and social *structure* – the rules and relationships that tie us to groups and organisations through, for example, education, work, leisure and family. Though we cannot present these debates fully here (see Giddens, 1984; Cohen, 1989; Parker, 2000), it is important to have some idea of the extent to which our capacity to make and interpret representations is a result of either ourselves as free-thinking creative individuals or comes from larger social and biological forces about which we have little understanding or control. Although positions have altered frequently in recent decades, much contemporary theory concerned with the production and interpretation of cultural materials has advocated an approach in which individual human agency and wider social forces, rules and structures both play a role.

Given that the current age seems to supply so many choices, the idea that people are merely the product of external forces seems difficult to sustain. Increasingly life in modern consumer society leads to the belief that people are free to create a distinctive and individual life for themselves from the vast range of consumer and lifestyle choices available. As Alan Warde (1994, 878; see Gronow, 1997, 5) remarked: 'In a world where there are an increasing number of commodities available to act as props in this process, identity becomes more than ever a matter of the personal selection of self-image. Increasingly, individuals are forced to choose their identities.' Just as people make and remake representational systems in society, they also express their attitudes and values through culture in preferences for everything from clothes, food, music, films and books to lifestyle, house, job and education. Yet this does not mean that these choices are made without constraints. The individual's interests, values and choices are a product, among many other things, of childhood, education, socio-economic status, nationality, the historical period in which we live, and economic, social and political structures. If cultural preferences are informed by who we are, where we come from and when we lived, then cultural products can tell us a great deal about how people think and feel at different times and places.

From this position, it would be easy to be drawn into a deterministic view of culture. Many historians and sociologists concerned with the 'action-structure' debate have argued that individual choice and creativity play an important role in producing and interpreting both cultural representations and, more generally, social relations. In this context, the historian E.P. Thompson (1978) likened social life to a 'game' where the pitch, regulations and general rules of behaviour are given, shared and understood before the players take to the pitch. Yet, players can then use their individual and collective abilities and imagination to outsmart the opposition. Though this analogy may seem crude, it is echoed in the work of sociologists. Anthony Giddens (1984), for example, argued that the structures, rules, regulations and material circumstances of life are both enabling and constraining. They provide a set of limitations within which to live and work while, at the same time, affording the materials with which to improvise, create and build something new – in the process transforming the original circumstances of our actions. The concept of 'habitus', derived from the work of the French sociologist Pierre Bourdieu (see Lane, 2000), is useful in this regard:

> Habitus represents the resources that people draw on to make activity happen, but at the same time limits potential. Thus it is the key mechanism that interweaves the creativity of individuals with their direct involvement in the reproduction of structural resources.
>
> (Layder, 1994, 156)

Thought of in this way, the making and interpretation of representations both reproduces and remakes, or creates anew, the social relations on which they initially depend. These ideas lead us to think about the role of representations in society as one that, on the one hand, takes account of structures and contexts and, on the other, pays due regard to creativity and transformation, negotiation, dialogue and contest between different sets of social groups with particular interests. The next sections of this chapter therefore explore these issues, taking as an example the representations produced as part of the debate about genetically modified organisms and their environmental consequences.

Representation and society

Exercise 3.4

Read the following report by John Vidal published in the *Guardian* on 21 September 2000. Under a headline that identified the article as one in a series of Special Reports on the Genetically Modified (GM) crops debate, the reporter discusses an environmental protest organised by Greenpeace against GM crops that took place early on the morning of 26 July 2000.

Think about the relationship between the *practical* and *symbolic* consequences of this action. By the term 'practical', we mean the physical impact of the action and by the term 'symbolic' we refer to the cultural meaning of the action.

Under the headings of practical and symbolic make a list of the resources and outcomes mobilised by this action. By the term 'resources' we are thinking of any material assistance, technology or equipment, skill, knowledge or social relationships used in the action. By the term 'outcomes' we are thinking of the intended and possibly unintended effects of the action.

Confrontation on GM battlefield

Greenpeace trial: short but epic meeting of farmers and protesters

When Lord Melchett, former Labour minister, working farmer and head of Greenpeace UK, arrived at remote Walnut Tree farm, near Lyng, Norfolk, with 27 others just before dawn on July 26 last year, he knew the environmental group had just a few minutes to destroy the six acre GM maize crop being grown by the three Brigham brothers for seed company Agrevo.

With luck, the farm machinery they had brought would work faultlessly, the police would be alerted too late and the 'polluting' maize would be bagged up and on its way back to the company's head office in King's Lynn before anyone could stop them.

The first half of the operation went to plan. Lord Melchett had set off at night from his family farm, some 30 miles away at Ringstead, Norfolk, with an industrial crop cutter on the back of a wagon. Most of the other Greenpeace members had travelled there from London in two minibuses.

The rendezvous was near the GM field at 5am. By the time Lord Melchett arrived, the 27 volunteers – who included nine Greenpeace staff and supporters from all over Britain including a Baptist minister – were waiting behind a hedge.

Decontamination suits

Within seconds, the padlock on the gate barring the GM trial field had been cut, lorry and cutting machine had gone in, and followed by the volunteers all dressed in white 'decontamination' suits. The gate was immediately re-padlocked and Melchett spent an infuriating few minutes trying to set up the machinery.

What Greenpeace had not counted on was the alertness and the anger of the Brigham brothers and their determination to protect their crop. Within seconds of the environment group's cutter speeding through the field like a whirling dervish, the three brothers were heading from the farmhouse towards the field on foot and by tractors.

The meeting was short but epic. There were moments of real drama and danger for the volunteers and the clash was described as a confrontation between outsiders and locals, with Greenpeace being alternately hailed by the anti-GM lobby and condemned by government and most of the media.

William Brigham tried to physically stop the volunteers who had immediately begun cutting the crop by hand and putting it into bags. Meanwhile, John and Eddie Brigham were turning their tractors into mobile battering rams.

The two Greenpeace minibuses were damaged. As one brother on his Massey Ferguson ambushed the Greenpeace cutter on its second revolution round the field, striking it down with one blow of his front end loader, another was slamming a heavy JCB-type shovel on top of the Greenpeace lorry preventing it moving.

Within 10 minutes, the Greenpeace action was effectively over, with only the volunteers left cutting by hand. At most, one sixth of the field that was about to flower and pollinate, causing what Lord Melchett said would be 'inevitable pollution', had been cut down.

The brothers had caused an estimated £5,000 of damage to the Greenpeace machinery and Greenpeace had caused about £650 of damage to the crop.

The police arrived at about 5.30am. By this time one of the Brighams was speeding round the field in his tractor in pursuit of anyone in a white suit and the volunteers were hiding or throwing themselves deep into the maize to avoid injury,

Ecological arguments

As the police moved in to arrest Greenpeace members and lead them to relative safety, the political and ecological arguments started. William Brigham confronted Lord Melchett, accusing him of being a criminal: 'I find it amazing that a man calling himself a democrat and is a former government minister sees fit to take the law into his own hands.'

Lord Melchett said: 'This is decontamination of the countryside. This crop shouldn't be grown. We are doing something that the public wants and is for the benefit of the environment.' He added that Greenpeace was trying to protect other farmers.

William Brigham argued that he, rather than Lord Melchett, was working for the community, and that he and his brothers were acting in an ecologically responsible way.

'I wanted to trial these crops to see if there were any downsides,' he said.

Within hours, the Greenpeace 28 had been driven off and Lord Melchett was preparing to spend two nights in Norwich prison after being refused bail.

In responding to this exercise, your answers might include some of those listed in Table 3.1.

Table 3.1 *GM protest: resources and outcomes*

	Practical	Symbolic
Resources	Farm machinery	Decontamination suits
	Two minibuses full of volunteers	Farming skills and equipment – a farmer (Lord Melchett) using an act of farming (reaping the crop) against the farming industry
	A knowledge of farming and how to use crop-cutting machinery, etc.	
	Bolt cutter	Lord Melchett and a 'Baptist minister' – authority figures participation in the protest
	Decontamination suits	
		Assertions of popular sympathy and moral right to engage in such action
Outcomes	Cutting down the GM crop that was about to flower and spread its pollen into non-GM fields	Lord Melchett and the 'Baptist minister' in prison
		Ritual destruction of crops

Table 3.1 *continued*

Practical	*Symbolic*
Destruction of scientific trial of GM crop, thus damaging the scientific programme on GM crops	Good publicity for Greenpeace and the anti-GM crop cause
Newspaper reports of the protest and any subsequent legal case	Adverse publicity for the seed producer
A legal test case concerning the legality of such protest action	Intimidation of farmers participating in trials

In light of the previous discussion, two things are readily apparent. First, this environmental protest showed the use of routine objects, knowledge and activities in a culturally creative manner. It shows how a range of resources that form part of the farmer's 'habitus' could be employed in a manner that was critical of everyday life and even lead to its transformation. Second, it shows the difficulty of trying to differentiate between the practical and the symbolic aspects of activity. In this instance, cutting the maize was both a practical action to remove it before it flowered and a symbolic action drawing media attention to the anti-GM campaign. Lord Melchett's skills in using farm machinery practically assisted the protest. At the same time, his status as a former government minister, peer of the realm and head of Greenpeace made it also highly symbolic.

The institutional context

Environmental representations reflect institutional context. All such representations are produced, transmitted and consumed within specific forms of social organisation. As such, social contexts shape their use, purpose and form.

Exercise 3.5

The ensuing text contains some brief quotations from the website for Monsanto, the agricultural chemical and technology giant who is a world leader in the development of genetically modified crops.

Why do you think Monsanto produces an 85-page report published through its website concerned with 'sustainable development'? Why does it feature so prominently on the Monsanto homepage?

---

continued

> Pick out some key words and phrases from the statements by Monsanto entitled 'vision' and 'principles' which show how language is used here to convey an impression that GM technology and Monsanto's role in its development is essentially benign?

Monsanto vision

Monsanto is a life sciences company addressing the food and health needs of a rapidly expanding world population while recognizing the importance of environmental sustainability. Through innovative technology and breakthrough products that link the fields of agriculture, nutrition and medicine, we are dedicated to helping people everywhere live longer, healthier lives.

Monsanto principles

As we move forward, we will:

- embrace safety and health in everything we do;
- protect and enhance the ecosystems of the planet;
- bring value to communities where we do business;
- act with care and respectability in developing and applying technology;
- create opportunities for diverse, talented people who want to make a difference;
- proceed with creativity, courage and speed to find breakthrough solutions to humankind's fundamental problems;
- link extraordinary financial success to contributing to the well being of people and the planet;
- keep our word, learn from our mistakes, value the ideas of others and report our progress.

Companies developing GM technology have generally received adverse publicity. This is partly due to worries about the creation of mutant strains, loss of environmental diversity and greater dependence on chemicals in farming and partly reflects concerns about the control that high-technology, multinational companies might exert by patenting crop strain variants, producing seeds containing 'terminator genes' and creating varieties that rely on the same company's patented chemical fertilisers, herbicides and pesticides. Many environmentalists fear that control of the 'seed-feed nexus' by a few companies will give them unprecedented dominance in global agriculture. Campaigners, therefore, often portray such companies as greedy and lacking in social conscience. By contrast, the companies concerned are keen to portray themselves as using technological advances to solve environmental problems in a socially responsible way, making the case for the benefits of GM technology in a calm and serious manner (see also Chapter 1).

Monsanto, for example, presents scientific evidence to justify environmental benefits of GM technology and farmers' willingness to embrace GM products. In addition to the 'vision' and 'principles' set out above, therefore, its substantial *1999/2000 Report on Sustainable Development* charted a range of environmental performance indicators for the company. These included the overall environmental impact of Monsanto's global operations measured in terms of energy efficiency, water consumption and production of greenhouse gases; and lists of projects that use Monsanto products to address environmental and social problems. The document clearly represented the company as environmentally responsible, socially accountable and centrally concerned with technologies that bring environmental and social benefits. The website's authors tried throughout to allay potential fears about GM technology and about Monsanto's corporate objectives. The idea of producing a report on 'sustainable development' let Monsanto speak with the voice of the environmental movement and suggested, at the very least, common purpose. The set of 'principles' outlined its intention to:

- protect and enhance ecosystems;
- act with care and responsibility;
- contribute to the well-being of the people and the planet;
- learn from our mistakes;
- report on progress.

This is designed to address the specific criticisms made by environmental groups and others about GM technology and corporate agribusiness.

The ways in which corporate organisations represent themselves extend beyond reports and images to the structure of the organisation itself. For example, these can take the form of:

- *Evidence of internal regulation.* For Monsanto, key environmental, safety, health and sustainable development officers oversee the implementation of 'eco-efficiency guidelines' developed by the World Business Council for Sustainable Development.
- *Evidence of philanthropic redistribution.* The Monsanto Fund is a charitable organisation which contributes to socially worthwhile causes in areas of the world in which Monsanto operates.
- *Corporate makeover.* Monsanto distances itself from problems in the past casting these as the responsibility of a previous regime. The 'Letter from the Chief Executive Officer', which introduces the *1999/2000 Report on Sustainable Development* endeavoured to represent Monsanto as a new company. It began by saying that: 'In the last two years we've been through some significant organisational changes here at Monsanto'. It concluded: 'Our intent for this first environmental and sustainable development report of the *new* Monsanto Company is to offer a statistical look at our operations and

a window into the impact of our products and people on the world' (emphasis added).

In this way Monsanto reinforced the words, statistics and images of its publicity with an organisational structure that represented the company's concern for 'sustainability'. Yet adopting a vocabulary and a set of structures organised around the idea of 'sustainable development' did not necessarily imply that the company shared a conception of 'sustainable development' with the environmental movement.

Exercise 3.6

Monsanto's web page pictures a quiet but clearly ordered and productive agricultural landscape, with crops growing luxuriantly in perfectly manicured rows reaching out to the horizon. In the distance, the observer sees a farmstead and to the right of the picture three people examine the growing crop. Though agricultural production seems to take place on an almost super-human scale, production is anchored to human effort, our needs and values through the symbols of distant farm and nearby figures.

How do you think environmentalists would interpret such an image of an agricultural landscape?

Re-read the statements by Monsanto entitled 'vision' and 'principles'. Given what you know about Monsanto's business interests, what kind of 'sustainable development' is implied by the image on the Monsanto homepage?

Environmentalists might interpret this image as a representation of all that is wrong with modern commercial agriculture. Rather than interpreting it as a well-farmed, efficient and productive landscape, they might point to the huge expanse of monoculture, total lack of hedgerows, woodland or other diverse habits. For many environmentalists, this image represents an archetypal example of the ecological deserts created by agribusiness. This all suggests that even though Monsanto sometimes uses language similar to that of the environmental movement, the aims and objectives which lie behind these words are very different. Your answer to the question concerning Monsanto's conception of 'sustainable development' might include some of the following:

- year-on-year increases in agricultural outputs;
- continued cycles of technical and scientific innovation;
- sustained levels of business activity and profit margins;
- current living standards and levels of consumption must be sustained;
- technological innovation is the best way of achieving this.

Environmental theorists recognise that the term 'sustainable development' can imply a wide variety of environmental, political, social and economic strategies (see also Chapter 8). Monsanto's 'principles', for example, reflect this ambiguity. Phrases such as 'embrace safety and health', 'bring value to communities', 'contributing to the well-being of people and planet', have many interpretations. Depending on one's values, 'planetary well-being' could suggest anything from the maintenance of current lifestyles to a radically non-commercial form of society founded on ecological principles. Some environmentalists now refer to such corporate adoption of the language of 'sustainability' as 'greenwash'!

Exercise 3.7

Figure 3.2 is a mailshot leaflet distributed on behalf of the environmental organisation Friends of the Earth in 1998. It concerns their campaign against the open air growing of GM crops in the UK and sets out in brief the group's concerns and objections to the unrestrained use of GM technology and the role of corporations in science and decision-making.

Comparing Monsanto with Friends of the Earth, think about the possible different interpretations of the terms; 'embrace safety and health', 'bring value to communities', 'contributing to the well-being of people and planet'.

In thinking about these matters, it is very useful to introduce the notion of 'ideology'.

Ideology

Although it has many different meanings 'ideology' is regarded here as a pervasive set of ideas, beliefs and images shared by people who belong to a specific group and used to make sense of the world around them. It is a concept that draws attention to the way in which the interests of particular social groups, organisations and institutions are expressed in their activities. 'Interests' here can be defined as the goals, aims and objectives of a particular group or organisation, whether these are consciously worked out or just tacitly assumed. While recognising other meanings of the term, Thompson (1984), preferred to treat ideology in this way, as 'a system of signification which facilitates the pursuit of particular interests' and which sustains specific 'relations of domination'. By using the term 'domination', he suggested that the expression of interests also furthers the interests of that particular group, organisation or institution against the interests of others with which it might conflict or compete.

IRREVERSIBLE

Once released into the
environment, experimental
gene crops are likely to
replicate and interbreed
with wild plants. This
immoral genetic pollution
will last for as long as
there is life on Earth.

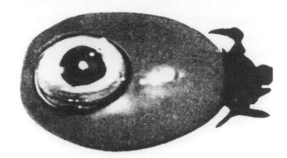

Take action now.

Join FOE.
Help us put a stop to
genetic food experiments.

➔

Genes are forever.

Genetically modified food crops give us no second chances. Once they are introduced into the environment, they're here to stay. They can interbreed with wild plants. The genetic make up of these plants will – for better or worse – be with us for as long as there is life on Earth.

Yet these crops are being tested now in Britain – in the open air. Some sites are as big as a hundred football pitches. The foods under test contain the characteristics and genes of unrelated plants, viruses and bacteria. The biotechnology companies have created sweetcorn that contains the anti-freeze gene of an Arctic fish. Potatoes have been modified with genes from soil bacteria to kill insects. No-one knows the long term consequences of these experiments.

Strict rules supposed to prevent genetically modified genes from entering the food chain apply to test sites in Britain. **But elsewhere genetically modified crops have already broken out of the test sites.** Research shows that they can interbreed with wild plants and form new superweeds resistant to weedkillers. Increasing amounts of herbicide will be needed to kill them. We could face a future where more weedkiller, and more toxic chemicals are sprayed on the land, not less.

The biotechnology giants are playing with the building blocks of life. **Open air genetically modified crop testing should end.** No genetically modified crops should be allowed into the environment, or into our food, until scientific research proves that this technology is safe.

Figure 3.2 *Mailshot, Friends of the Earth, 1998. Courtesy of Friends of the Earth*

UNNECESSARY

No-one needs genetically modified food, except the companies investing in it. The only reason for its existence is to increase agri-business profits.

Take action now.

Join FOE.
Help us put a stop to genetic food experiments.

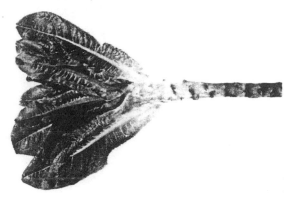

Modify me! ➜
Find my other half.

Just what are gene scientists playing at?

Genetically modified food is dangerous because its impact is unpredictable.
Genes work in infinitely complex ways that scientists are only just beginning to understand. What is known is that genes are the basis of all life. They carry the delicately coded information from which proteins create every structure and function of every living thing, from bacteria to humans.

Genetic engineering cuts genes from one species, and inserts them into another. **Things are made which no natural process would produce.** By mixing the genes of different plants and organisms we may unwittingly destroy nature's own genetic checks and balances.

The dangers to human health are real. **The deaths of 37 people in the United States are thought to be linked to a food supplement called Tryptophan, made from genetically modified bacteria.** A further 1,500 people have been permanently disabled. Genetic experiments on soya containing a brazil nut gene had to be stopped when it was discovered that the soya beans caused a reaction in people allergic to nuts.

Gene manipulation is a new technology. Scientists do not fully understand it. No-one can predict its results. **We should learn the lessons of the past, like BSE, and stop genetic food experiments now.**

The representations of Monsanto and Friends of the Earth certainly revealed evidence of the competing interests typical of an ideology. One might easily interpret these as the interests of these organisations and as overt expressions of their ideologies but, as the example of Monsanto shows, stated aims and objectives may only tell part of the story. As with Turner and Defoe, it is important to set representations in context. With regard to Monsanto, their specific concerns with 'sustainable development' need to be viewed alongside the company's commercial and other activities. Environmental activists have established groups such as Corporate Watch, with the expressed intention of scrutinising the activities of large-scale business in this regard. Given the type of evidence they often supply, the term 'interests' often implies rather more than the company's stated aims and objectives. That applies even, or perhaps especially, when the organisation itself wishes to give the opposite impression by producing formal statements of 'vision' and 'principles'.

Statements by Friends of the Earth need similar scrutiny. Leaders may purport to speak for the organisation, but not all members will necessarily share the organisation's views in their entirety. The opinions expressed on the leaflet are almost inevitably the result of considerable discussion, debate and dissension. The opinions of members of Friends of the Earth could range from support for limitation and regulation of GM crops through to demanding an immediate and permanent ban on this technology. Evidence on this issue might be found, for example, in the records of meetings within the organisation, correspondence in newsletters and periodicals, other campaigns, the activities of members or groups of members in specific localities and, if available, in earlier or unpublished drafts of the leaflet itself.

These are good reasons, then, for maintaining an approach to culture through concepts like 'habitus' that recognise the cultural significance of everyday actions and routines that are not necessarily formalised in terms of stated aims and objectives. We argue that this is the case even when representations, in the conventional sense of texts and pictures, are the primary objects of study. By finding meaning through a comparison of what people say they are doing and what they actually do, representations can be better connected to the broader social meanings and sets of environmental attitudes and values that are of interest here.

Classifying institutional contexts

It is convenient, though something of an oversimplification, to divide institutional contexts into those of production, consumption and communication (where the term 'institution' applies equally well to formal and informal social organisations):

- *Institutions of production* include the mass media, broadcasters, publishers, the arts, educational and scientific research organisations, central and local government, political parties, campaigning and pressure groups, advertising agencies and commercial and business operations.
- *Institutions of consumption* include commercial trading outlets, shops, and malls, theatres, concert halls, cinemas, consumer interest organisations, social groupings and sub-groupings, the 'family' and 'teenagers'.
- *Institutions of communication* link production and consumption and include radio, TV, drama, concerts, newspapers, books, reports, photographs, CDs, advertisements and the Internet.

Perhaps most important to remember at this stage are:

- All who produce cultural representations have 'viewpoints'.
- No cultural representation can exist without offering its consumers a position or 'point of view' to adopt.
- The 'viewpoints' of a particular group, organisation, or institution are an expression of its interests.

However, audiences take meanings and make sense of environmental representations in accordance with their existing knowledge, understanding, motivations, aims and objectives. Thus, interpretation of a representation is never just a simple reflection of the interests of either the producer or consumer (Branston and Stafford, 1996).

Exercise 3.8

The statement of Monsanto's vision and principles and the illustration in Figure 3.2 convey very different sets of interests vested in the GM debate. How do the representations from these different sources reflect the contrasting interests of production and consumption?

Can you itemise these in terms of underlying ideology?

On this matter, at the most general level the interests of Friends of the Earth are based in issues of consumption and consumers. This is expressed specifically in terms of food quality and, more generally, resource conservation. At the same time, Monsanto's interests have a basis in production and producers, whether defined in terms of agricultural output, the efficiency of agribusiness or Monsanto's own commercial performance. Some of the underlying principles behind these two perspectives may be characterised as follows:

Monsanto

- free enterprise;
- individualism;
- technocentrism (faith in the ability of technology to solve problems and improve life).

Friends of the Earth

- precautionary principle (concerning not undertaking actions when unsure about the possible outcomes);
- collective responsibility;
- social and biological diversity.

Multiple meanings

Representations, as will now be increasingly clear, may have multiple meanings, but not all meanings are equal. If interpretation of existing environmental representations is a creative act, and even if words and concepts shift their meanings (see above), then all environmental representations are to some degree open to a multiplicity of meanings both *actually* (at any one given time and place) and *potentially* (as and when they are interpreted and reinterpreted in the future). Media theorists (e.g. Morley, 1992; Branston and Stafford, 1996) have suggested that meaning varies with how we read 'representations', with differences between:

- a 'dominant' reading, where the producers' intended or 'preferred' meaning is entirely accepted;
- a 'negotiated' reading, where aspects of the message are accepted and others rejected;
- an 'oppositional' reading, where the fundamental premise of the message is rejected outright.

This is a useful classification. In the study of environmental representations, or indeed any form of cultural representation, it is widely accepted that the establishment of dominant meanings is a matter of power. At first hearing, for example, the assertion that we are subject to the power of others as we sit watching the television sounds faintly ridiculous. Yet in spite of the many channels available, television still constrains our choices, shaping our tastes and expectations. This, at least partial moulding of the viewer's experience is an expression of power. More generally, the power of the medium is revealed in what gets said and what never receives airtime. This may result from conscious censorship, for example, as when a government or its agents intervene to block the broadcast of certain material 'in the public interest' or in the 'interests of security'. Alternatively, it may be connected to the hidden social values that underpin commercial decisions, which simply claim that certain subjects will fail to interest the 'majority' of an audience.

In the case of Monsanto, we have already seen how the company tries to ensure a positive dominant reading of its activities in the way it represents itself to the world. It sponsors research and targets suitable worthy causes. As major players in the biotechnology industry, Monsanto also adopts a variety of legal and other methods to protect its interests. For example, it has won legal cases against other biotechnology companies such as Mycogen concerning access to gene technology (*Guardian*, 30 June 2000). The company also lobbies to influence national govern-ments and international agencies, with some measure of success. In September 2000 the US government, supporting the interests of multinational GM crop-producing companies, threatened sanctions on European exports if supermarkets in the European Union labelled products as containing GM foodstuffs (*Guardian*, 20 September 2000). As such, imposed anonymity of origin might overcome public resistance to GM foods.

Looked at another way, however, these actions to suppress information concerning GM foods in the political realm perhaps show how far environmental activists have succeeded in targeting GM foods as being environmentally hazardous. One particularly effective strategy has been to appropriate the metaphor of 'Frankenstein foods' in order to make a political point.

Exercise 3.9

Look at the examples of Frankenstein food imagery (Figure 3.3).

What impact do you think these images are designed to have? How does this compare with the calm and collected approach adopted by Monsanto?

The metaphor of Frankenstein foods directly relates to the figure of the megalo-maniac scientist whose single-minded obsession brought catastrophe. The images appeal to environmental campaigners due to their familiarity from countless film, television and comic strip adaptations. Indeed the visual image is sufficiently strong to need little textual explanation. As a moral message, it suggests that playing with the idea of life itself merely results in abject misery and death. More broadly, the metaphor plays on popular fears of monsters and mythical beasts that have been central to European folk culture since the Middle Ages (Warner, 1998). First published in 1818, the novel *Frankenstein, or the Modern Prometheus* by Mary Shelley was grounded in the environmental values and scientific thought of its day

Figure 3.3 *Frankenstein food imagery*

Sources: *Private Eye* **magazine, copyright Pressdram Ltd, 1999, and Press Association;** *The Week* **magazine**

(Shelley, 1818). At one level, it can be interpreted in a way that is sympathetic to the activities of science and scientists, even if that seldom forms part of popular mythology. As used in the GM food debate, the image is sometimes deployed humorously and playfully, sometimes seriously and shockingly, to gain attention. In this regard environmental campaigners have strategies open to them that are not available to the biotechnology industry, for which shock imagery might convey what they would regard as the wrong message about their activities.

In discussing this subject, however, we are confronted by the arena of 'cultural politics'.

Cultural politics

Environmental representations, as noted previously, have political as well as cultural dimensions and the meeting point for these dimensions is the field of cultural politics. When asking questions about representations, it is often important to know:

- Who has the power to say what about whom, when and where?
- Who has the power to persuade and convince others?
- Whose culture is the official one and whose is subordinated?
- Whose views and values encapsulated in representations are accepted as valid or true? and whose are marginalised?

These types of questions concern *cultural politics*, defined as the field concerned with the power behind meaning and the way that the exercise of such power advances the position of particular groups and their interests (Barker, 2000, 383). Struggles over environmental issues can trigger deep emotions, with ready expression in the cultural politics of the groups involved. These might include contests over meaning, identity and sense of self and be expressed in the form of nationalism, territoriality, xenophobia, racism and sexual chauvinism. Representations of environmental degradation and exploitation can also serve as a rallying cry for marginalised and oppressed groups – especially where the areas concerned are ones for which they have deep attachment. In such instances, outrage at expropriation and abuse can assist the construction of new and resistant identities. In turn, those identities can contribute to a wider political struggle to transform society (after Jordan and Weedon, 1995, 4–6).

Chapter 1 suggested that representation could be simultaneously a cultural and a political act. Representations, as we saw there, both depict (*resemble*) and stand in place of something or someone (*represent*). They define, but they also speak for that something or someone. In this sense, all acts of cultural representation might be regarded as political because they are fundamental to the activities by which differing sets of individual and group interests compete to be heard within society. Analysing the power to define what things mean, to win support for certain kinds of cultural values and identity over others, and to examine how this in turn shapes and is shaped by access to wealth and social resources, are central to understanding cultural politics. Certainly for the study of environmental representations, the issues of power encapsulated in the idea of cultural politics are crucial to our ability to make connections between culture and the expression of social values concerning the environment. Following Jordan and Weedon (2000, 13), we might highlight several key areas:

- *The power to name*: concerning both the assertion of identity and claims to represent reality that are central for the struggle to gain support for certain kinds of cultural value in environmental debates.
- *The power to represent commonsense*: implying the power to embody and articulate 'our' basic notions and values. These include values concerning our relationship to the environment, its use and misuse, what is worth conserving and what we can legitimately exploit.
- *The power to create 'official versions'*: that is, to give authoritative accounts of things. This is largely a property of institutions, but it affects all areas of individual and social life. It includes, for example, government statements concerning such subjects as road building, nuclear power and waste disposal. It would cover the work of scientists producing reports on subjects like global warming, habitat loss or biotechnology. It also includes news reports of droughts and famines, as well as ethnographic accounts of threatened ways of life in the developing world.
- *The power to represent the legitimate social world*: implying the power to speak on behalf of 'respectable, decent society'. Closely allied to the power to represent commonsense, it also implies the ability of environmental groups, conservation organisations and others to express alternative futures for the well-being of the planet. By harnessing the idea that ordinary people are excluded from decision-making critical to their future, conservation groups assume the right to act on their behalf.

Strategies to enforce dominant readings through the exercise of power include the following:

- *Coercion*: including regulations that define what can be said in public (e.g. censorship) and rules which limit access to cultural institutions and media of communication through, for example, qualifications for age, gender, education, economic resources, or nationality.
- *Persuasion*: including strategies that seek to justify a representation or interpretation by making a series of claims for its truth. This might be done, for example, by asserting that the maker of the claim is a disinterested party merely conveying the simple truth; or by associating a claim with an individual, institution or place that will be recognised by the audience as being authoritative.
- *Negotiation*: comprising strategies that recognise the need for co-operation and consent between interest groups for any attempt to achieve wider ideological acceptance. Here particular interest groups recognise their mutual

weaknesses as political agents, making compromises in the expression of values and ideologies and suppressing differences.

The first two sets of strategies imply dominance of both social structure (rules and regulations) and producer-fixed meanings (persuasion). Yet, as noted earlier, though important these form only part of the story. The exercise of cultural dominance must always incorporate some form of negotiation even if coercion and persuasion are still central. The concept of 'hegemony' has value here.

Hegemony

Derived from the Greek word *hegemon* meaning 'leader' or 'ruler', hegemony describes the way that politically powerful groups maintain their power by presenting their view of the world as 'commonsense' and part of the natural order for those who are in fact subordinated by it (Bullock, 1988, 379). The Italian Marxist Antonio Gramsci (e.g. 1971) supplied one plausible explanation of how this process works. According to Gramsci, since power in bourgeois democracy is as much a matter of persuasion and consent as it is force, any dominant group has to a greater or lesser degree to acknowledge the existence of those who it dominates by winning the consent of competing or marginalised groups in society. Hall (1997) argued that unlike the 'fixed grip' over society implied by 'domination', 'hegemony' is won in the to-and-fro of *negotiation* between competing social, political and ideological forces through which power is contested, shifted or reformed. Furthermore he asserts that *representation* is a key site in such struggle, since the power of definition is a major source of hegemony:

> The concepts of hegemony and negotiation enable us to rethink the real and representational in a way that avoids the model of a fixed reality or fixed sets of codes for representing it. And they enable us to conceptualise the production of definitions and identities by the media industries in a way that acknowledges both the unequal power relations involved in the struggle and at the same time the space for negotiation and resistance from subordinated groups.

The next exercise provides an opportunity to see what these ideas mean in practice.

Exercise 3.10

Return to our discussion of the GM crops debate and the examples from Greenpeace, Monsanto and Friends of the Earth above.

List how corporate and environmentalist perspectives differ on the issue of GM crops with respect to:

- the power to name;
- the representation of commonsense;
- the creation of 'official versions';
- the representation of the legitimate world.

Your answer may include some of the following:

- *The power to name.* One important way in which the perspectives contrast concerns the naming of crops as either safe or as pollutants. Greenpeace justified their attack on the field of maize by saying that to remove the crop was to get rid of 'pollution'. In court Lord Melchett took this analogy further by likening GM crops to nuclear waste. By contrast, Monsanto make much of safety issues in their publicity.
- *The representation of commonsense.* This can be found in two seemingly opposing views. On the one hand, there is the environmentalist focus on the precautionary principle, found in the assertion that commonsense decrees not undertaking any action before there is complete confidence that the outcome will not be harmful. On the other hand, there is expression of corporate faith in the ability of technology to find 'solutions to humankind's fundamental problems'. The biotechnology industry would probably say that this is perfectly compatible with the precautionary principle and that technology is fundamental to avoiding any adverse consequences of future developments.
- *The creation of 'official versions'*: for example, in the way that both environmentalists and the biotechnology industry make extensive use of scientific and social scientific research to back up their claims with authoritative accounts. Much information used in the preparation of these pages has been found at the *Guardian* newspaper 'GM debate: special report' website. This, therefore, constitutes another form of official version. Although ostensibly dispassionate, the *Guardian*'s left-of-centre sympathies were noted earlier (see Chapter 2).
- *The representation of the legitimate social world*: as exemplified by the differing scope and scale of social responsibility assumed by corporate and environmentalist interests. The legitimate social world for corporate interests is based in commerce, markets, profit margins and shareholders' accounts;

it rests firmly in direct human concerns. This is suggested by the Monsanto principle to find 'solutions to *humankind's* fundamental problems'. No mention is made here of the broader environmental issues, even though the wording implies involvement with welfare and the greater good. By contrast, environmentalists would argue that everyone is responsible for the environment in its widest sense. The legitimate social world, therefore, includes the wider natural world and the welfare of those things for which we are unlikely to receive direct financial reward.

To round off this section on multiple meanings, we shift the focus away from Monsanto to a notable episode in the debate over GM foods.

Exercise 3.11

Read the following account from the article published in the *New Scientist* on 20 February 1999.

While thinking about the four key powers involved in the exercise of cultural politics, find examples of coercion, persuasion and negotiation in the story of Arpad Pusztai.

What specifically hegemonic move does the British government make?

Anatomy of a food scare

10 August 1998: Arpad Pusztai, a biochemist at the Rowett Research Institute in Aberdeen, appears in a documentary on British TV to warn about the inadequate testing of GM foods. He claims to have carried out experiments showing that feeding genetically engineered potatoes to young rats suppresses their immune response and harms their growth and development. Opponents of GM foods everywhere seize upon Pusztai's remarks.

12 August 1998: The Rowett says Pusztai muddled his results and was wrong to talk about unpublished findings. The institute says that the rats had eaten not GM potatoes but ordinary potatoes spiked with a jack bean lectin, one of a family of proteins used by plants to ward off insect pests and which are often toxic to mammals. Environmentalists cry 'cover-up'. Biotechnologists say it is much ado about nothing, as nobody planned to sell potatoes engineered to carry the lectin and in any case, even if Pusztai had done the experiments he described, all it proved was that if you insert a gene for a toxin into a potato the potato becomes toxic.

14 August 1998: Pusztai, who is past retirement age, is suspended and told that his annual contract will not be renewed. He is instructed not to speak to the media about his results. The gagging order remains in place to this day.

28 October 1998: A panel set up by Philip James, director of the Rowett Research Institute, criticises Pusztai, but does not accuse him of scientific fraud. It becomes clear that, despite the Institute's initial claims, his experiments did involve potatoes engineered to contain a gene for a lectin. However, the Institute claims there is no 'statistically significant' evidence in Pusztai's data suggesting that the transgenic potatoes harmed rats. Its report into the affair also states that Pusztai's experiments focused on potatoes containing a lectin from the snowdrop, not from the jack bean, as was previously thought.

November 1998: Pusztai's supporters circulate his 'alternative report' among sympathetic scientists. This contradicts the Rowett's report, reiterating the claim that GM potatoes harmed the rats.

12 February 1998: Twenty scientists from 14 countries who have examined Pusztai's report accuse the Rowett of bowing to political pressure. The group, including former associates of Pusztai and active opponents of biotechnology, calls for a moratorium on GM crops on the grounds that the study reveals an unforeseen hazard that would not be picked up by standard toxicity tests. Environmental groups say this is the first evidence of toxicity caused by the process of genetic engineering itself.

13 February 1999: The British government reflects calls for a moratorium amid allegations that it is in the pocket of the biotechnology industry.

14 February 1999: Further claims of a cover-up surface when it is revealed that the Rowett received £140,000 of funding from GM food giant Monsanto before the Pusztai affair blew up. Newspaper reports also claim that the British government offered millions of pounds in inducements to encourage biotechnology firms to invest in Britain.

15 February 1999: Members of the government show signs of bowing to media and public pressure. Environment Minister Michael Meacher says he will bring wildlife specialists onto the committee that oversees the release of GM crops into the environment, which has come under fire for being dominated by scientists associated with the biotechnology industry. He also floats the idea of establishing a 'GM food commission'. Meanwhile, opposition politicians call for the resignation of Science Minister Lord Sainsbury, who in the past has invested in companies with interests in GM products.

The story of Arpad Pusztai vividly shows what can be at stake in struggles to represent environmental issues such as those that concern GM technology. It shows the vital role the media play in the representation of environmental stories and events, even those located in worlds of factual evidence and 'hard' scientific truths. Governments react to public opinion driven by media coverage. Environmentalists vociferously champion the cause of scientific research that appears to support their case. Scientific establishments try to minimise the adverse publicity caused by research findings when they might interfere with income from big business. A scientist loses his job and has a lifetime's reputation of good work threatened by outside interests. There are many examples of coercion, persuasion and negotiation in this story:

1   *Coercion*: Arpad Pusztai was told not to speak to the press and lost his job.
2   *Persuasion*: biotechnologists said it was much ado about nothing. Nobody planned to sell potatoes engineered to carry lectin and all Pusztai's experiments showed were that if you insert a gene for a toxin into a potato, then the potato becomes toxic.
3   *Negotiation*: bowing to public pressure the British government floated the idea of a 'GM food commission'.
4   *Hegemony*: the move by the British government was to involve wildlife specialists on the committee that oversees the release of GM crops into the environment. By this move, the government brought a group critical of government policy into the arena of policy making without necessarily giving them any special powers on the committee. The government thereby attempted to co-opt them as individuals, and perhaps also some of the environmentalists who listened to what they said, to the dominant viewpoint.

Discourse and social meaning

During this chapter we have moved from considering specific representations in isolation towards the examination of a number of representations together. Focusing on a single issue, in this case the GM crops debate, these representations have taken a variety of forms including visual images, text, scientific reports, newspaper articles, websites, corporate mission statements and campaigning mailshots. In each case we have stressed the specific qualities of each mode of representation, while also trying to show how these can be used together when examining some of the environmental attitudes and values that inform the debate. We have shown how the production of meaning is bound up with the expression of power and is always an issue for contestation through the mechanisms of cultural politics.

Two further notions that now need adding to the analysis of the production of environmental representations are the concepts of 'discourse' and 'hybridity'.

Discourse and hybridity

The word 'discourse' is in fairly common use, although alternatives such as 'discussion' or 'conversation' are more likely in everyday speech. The *Oxford English Dictionary* (iv, 430) identifies the word primarily with speech or writing, in that it includes the following:

- to speak or write at length on a subject;
- to pass from premises to conclusions, to reason;
- communication of thought by speech, 'mutual intercourse of language'.

Putting these together, popular use of the word 'discourse' suggests structured, and often lengthy communication with verbal utterances or written material of greater magnitude than the sentence. It also suggests complex utterances by more than one speaker or writer. More frequently, discourse suggests the turn-taking interaction between two or more, as well as the linguistic rules and conventions that are brought into play to governing such interaction in their given context. To be effective, people need to speak the same language and it is helped if they share the same assumptions and values. Groups of people who communicate with these shared qualities are called discourse communities (WSU, 2003).

So far, this identifies discourse with language, but cultural theorists usually have something broader in mind when they use the word. The French cultural theorist and historian Michel Foucault (e.g. 1970), for example, argued that discourse was not purely a 'linguistic' concept, but concerns language *and* practice; in other words, both what one *says* and *does*. A discourse is a *regulated* way of speaking or writing about the world, where 'regulated' means that it takes place according to the rules and conventions of the society and its culture. A discourse then is not only concerned with what might be said or written about something, but also includes all the added rules, conventions and ideas that support those verbal or written statements. It always involves the symbolic and the practical and we will refer to *discursive practices* at various points in the ensuing chapters ('discursive' being the adjective from 'discourse'). Discourses generate 'knowledge' about particular objects or concepts because:

- they are coherent: they offer a lucid and consistent account of things;
- they are self-referential: they contain a body of statements that are mutually supporting and not contradictory;
- they offer an account of reality;

- they shape the rules of what can be said and known about these objects or concepts, since each discourse is distinguished by its own rules for indicating what is true, real, good, legitimate or acceptable.

There are various ways to classify discourses. For example, they may be formal or informal:

Formal discourses relate to those regulated ways of speaking or writing about the world found in the realms of government, law or education; or specialist areas such as biology, economics or landscape painting. Discourse in biology, for instance, is regulated by many forms of professional rules and conventions besides the normal rules of scientific inquiry and experimental procedure. These include the conduct of peer-review in journals, conference organisers deciding what papers to accept or which to reject, the deliberations and decisions of job interview panels, and communicated decisions of funding bodies in deciding which projects to support.

Informal discourses relate to those regulated ways of speaking or writing about the world that are not systematically codified as formal bodies of knowledge or principles of practice (as they are in law and economics). They include many systems of knowledge that are not necessarily accorded any special or legitimate place in society and are often accepted as forms of commonsense – although they may be codified as formal discourses elsewhere. Appropriate examples might include farming practices, formalised as agricultural science, or the activity of watching television, which becomes formalised as part of academic media research.

Three further ideas are useful:

The relationship of knowledge and power. Foucault argued that not only is knowledge always a form of power, but power is implicated in the questions of whether and in what circumstances knowledge is or is not to be applied. He thought that this question of the application and *effectiveness* of power and knowledge was more important than the question of truth (Hall, 1997, 48–9). In this sense, the ability to employ a discourse reflects a command of knowledge or set of skills. It also implies that this facility is employed in relation to people who lack such command and have no legitimate claim to such knowledge. For instance, command of a particular discourse, such as that of engineering or forestry, also allows control over those who do not have that command, such as displaced farmers or forest dwellers (Layder, 1994, 97).

The relationship between subjects and subjectivity. Discourses are more than ways of giving meaning to the world. Rather, they also imply forms of social organisation and practices that structure institutions and constitute individuals as thinking, feeling and acting subjects. Much of the previous discussion of cultural politics concerned issues of *subjectivity* – the power to define oneself or others. The concept of discourse makes an important contribution in this regard. It seeks to understand how a range of knowledges and competencies that enable us to function in society, to greater or lesser extent, *structure us* as individuals. Issues of subjectivity are important for the politics of subordinate groups in society and for the conduct of environmental action.

'Hybridity' by definition, is the state where different species or types are interbred to produce new strains or variants. Combining the *practical* and the *symbolic*, the material and the social, *discourse* stresses the *hybrid* qualities of social life. Hybridity draws attention to the ways in which the physical and social worlds interact and combine in a multiplicity of ways. The term is widely used, for example, in studies of GM foods, human, plant and animal biotechnologies. Such studies are frequently concerned with the *hybrid* nature of contemporary social life as it is formed in interaction between, humans, technology and nature (see Whatmore, 2002). It has been widely claimed, for example, that nature has become increasingly socialised to the point where all natural systems have fallen under the influence of human actions. This is summarised in the argument that 'first nature', or pristine nature, has become increasingly replaced by 'second nature', nature moulded or modified by human action, or even 'third nature', technologically simulated and synthetic environments (McKibbin, 1989; Soper, 1995; Macnaghten and Urry, 1998; Castree and Braun, 2001).

Returning once again to the GM foods debate, it is clear that much of the above discussion is centrally informed by the concept of discourse. The next, and final exercise seeks to make matters more explicit.

Exercise 3.12

Using the headings 'formal discourse' and 'informal discourse', make a list giving examples of each from our discussion of the GM foods debate.

Where do issues of knowledge and power, subjects and subjectivity, and hybridity feature in the GM foods debate?

Among many other possibilities, your answers might include some of the following:

- *Formal discourse*: research on sustainable development, the law on patenting gene technology, gene biochemistry, agricultural science, ecology.
- *Informal discourse*: Frankenstein foods, the *Guardian*'s GM debate, consumer power, responsible farmers.
- *Knowledge and power*: we might highlight the ability of biotechnology corporations to commission scientific research, influence governments in terms of food labelling and suppress research findings that are not favourable to them.
- *Subjects and subjectivity*: Arpad Pusztai being criticised as a scientist, dismissed from his place of work and not allowed to defend his position by speaking to the press.
- *Hybridity*: the genetically engineered potatoes in which the scientific manipulation of the natural gene structure has created a form of *socialised* nature. By this, we mean that these potatoes have been so modified by human technology for human purposes that they even rely on people for their reproduction.

Conclusion

This chapter has taken a series of historical and contemporary case studies to show some ways in which environmental representations connect with social activities and practices. It has argued that environmental representations cannot be considered in isolation but rather should be studied as part of wider patterns of discourse. To conclude, the following points should prove useful:

- All representations are open to a multiplicity of meanings.
- Audiences always play an active part in the construction of meanings.
- Meanings can change over time and must be studied in their historic context.
- Representations reflect institutional contexts and the interests of their producers.
- Representations are never purely a matter of images and symbols. They always imply at the very least access to resources, as well as some practical and material outcomes.
- Meanings are contested and reflect the differential powers available to groups and individuals.
- All natural systems are effectively socialised and therefore hybrid in character.

In the next three chapters in particular, the concept of landscape will be developed as a means of studying environmental representations as both *discursively* constructed and *hybrid* in nature.

Further reading

A provocative and ground-breaking introduction to the relationships between painting and society in historical context is:

John Berger (1972) *Ways of Seeing*, London: British Broadcasting Corporation and Penguin.

See also:

Peter Burke (2001) *Eyewitnessing: the uses of images as historical evidence*, London: Reaktion Press.
Richard Leppert (1996) *Art and the Committed Eye: the cultural functions of imagery*, Boulder, CO: Westview Press.

On using cultural materials to reconstruct historical environments, see:

Janet M. Hooke and Roger J.P. Kain (1982) *Historical Change in the Physical Environment: a guide to sources and techniques*, London: Butterworth Scientific.

For an introduction to studying society through culture, see:

Judy Giles and Tim Middleton (1999) *Studying Culture: a practical introduction*, Oxford: Blackwell.
Glenn Jordan and Chris Weedon (1995) *Cultural Politics: class, gender, race and the postmodern world*, Oxford: Blackwell.
Ziauddin Sardar and Borin van Loon (1997) *Cultural Studies for Beginners*, Cambridge: Icon Books.
John Storey (1993) *An Introductory Guide to Cultural Theory and Popular Culture*, Hemel Hempstead: Harvester Wheatsheaf.

For anthropological approaches to the relationship between environment and culture, see:

Philippe Descola and Gisli Palsson (1996) *Nature and Society: anthropological perspectives*, London: Routledge.
Kay Milton (1996) *Environmentalism and Cultural Theory: exploring the role of anthropology in environmental discourse*, London: Routledge.

On the relationship between environment and the mass media, see:

Derek Bousé (2000) *Wildlife Films*, Philadelphia: University of Pennsylvania Press.
John A. Hannigan (1995) *Environmental sociology: a social constructionist perspective*, London: Routledge.
Rom Harré, Jens Brockmeier and Peter Muhlhausler (1999) *Greenspeak: A study of environmental discourse*, London: Sage.

For methods, techniques and the relationship between the mass media and society more generally, see:

Gill Branston and Roy Stafford (1996) *The Media Student's Book*, London: Routledge.
Keith Selby and Ron Cowdery (1995) *How to Study Television*, Basingstoke: Macmillan.

4 The Classical, Medieval and Renaissance legacies

This chapter, the first of three on nature, landscape and the history of environment:

- introduces the historical study of environmental representations through the concepts of nature and landscape;
- traces the origins of Western environmental values in Classical, Medieval and Renaissance approaches to nature and landscape;
- highlights some key themes from these periods relevant to present-day environmental values.

The place of landscape

It is difficult now to imagine a way of understanding interrelationships between plants, animals, earth, climate and humans without using the word *environment*. 'Environment' draws simultaneously on a range of ideas, knowledge and other terms like 'ecosystem', 'community' and 'biodiversity', making it central to a wide range of discourses.

Exercise 4.1

Try to define our relationship with the world around us *without* using the term 'environment'?

What other words and concepts might perform a similar function?

It is probable that you found the words *nature* and *landscape* among the most useful alternatives to the term environment. Yet neither has quite the same emphasis or associations as 'environment':

- *Nature* has strong connotations of the physical world and the fundamental or essential aspects of that world. It is less useful for supplying a sense of the world as a set of interdependencies between Earth, animals, plants, climate and humans.
- *Landscape*, like environment, implies context or background. It can refer simply to the physical world, and gives a sense of the interaction between humans and the material world. Like environment and unlike nature, it may refer to the fusion of social and physical worlds in which all humans live. However, landscape suggests particular aesthetic and designed qualities that are not necessarily present in the term environment. In common with nature and unlike environment, landscape is a term with a very long history in Western culture.

Yet despite its familiarity at the present day, the word 'environment' is of recent origin, with few direct references to the 'environment' dating before the middle of the nineteenth century. This does not mean that societies were unaware of the environment; rather that they used different words to describe their relationship to it. Before the development of the science of ecology and the modern environmental movement, questions of human–environment relationships were most likely found in visual and writing sources relating to either *nature* or *landscape*. As Arnold (1996, 9) remarked:

> There is nothing new about the idea that the destiny of human beings is intimately bound up with the natural world. But what constitutes 'nature', what effect it has exercised on history, how far history can be written from a biological rather than social or cultural perspective, and what place the environment should occupy in the conceptualisation of historical time and space – these are issues that have long been debated and are still very far from being resolved.

Arnold chose the word 'nature' as key to understanding the history of society–environment relationships. More than this, he suggested that what is considered to constitute nature remains debatable. As shown in the GM debate in Chapter 3, scientists, environmental activists, commercial interests and the general public have different ideas about what is and what is not natural. The distinctive qualities of the concept 'nature' and the ways in which its meanings have built up and changed over time, therefore, are crucial to understanding its power in present-day society.

Nature

The earliest roots of the word 'nature' come from the Latin past participle *nazi* – to be born, from which we also derive *nation, native* and *innate*. In its earliest sense, nature was the essential character of something. Today the term can refer to any of the following, individually or in almost any combination:

- *An object*: a particular place or location, for example, wilderness, garden, countryside; the material realities of earth, flora and fauna; the measurable quantities of scientific description. Nature is physical reality.
- *A place*: a particular place or location, for example, wilderness, garden, or countryside. Nature is somewhere set apart from the human world.
- *An essence*: something intrinsic, a specific and defining quality of an object, an ideal type or pure form, the spiritual and immanent properties of religion, philosophy and pure science. Nature is something deep and spiritual, or abstract and philosophical.
- *A process or set of processes*: the emergent properties of growth, reproduction, the dynamic structures and systemic flows of ecosystems or biotic communities, the functional characteristics of scientific analysis. Nature is life itself.

From the above definitions, it is easy to see the powerful role that the idea of nature plays in the human imagination, providing us with facts (as object), peace and security (as place) and beliefs (as essence). According to Alexander Wilson (1992, 87): 'the culture of nature – the ways we think, teach, talk about and construct the natural world – is as important a terrain for struggle as the land itself'. The distinctive *cultural politics* (see Chapter 3) of nature are expressed in the ways that the term is used. For example:

- *Nature as an authority*: nature is the object of scientific study and therefore the basis of scientific truth. It is also a source of religious or spiritual truth, to the extent that people argue that nature expresses the will and demonstrates the works of God. For almost all major belief systems, to say that something is 'natural' is to say that it is *true*; that is, it is objective, factual or real. Reflection on nature as a basis for truth encourages us to imitate or project our representations of its best qualities back on to our own actions and creations. We see in nature what is *Good*; the right, best or proper course of action and what is *Beautiful*, that which is physically pleasing. Together *Goodness, Truth* and *Beauty* are extremely persuasive concepts and lie at the centre of modern value systems. Nature is consequently a powerful source for justifying arguments about how the world is now or should be in the future.

- *Nature as a model for human society*: modern conceptions of nature inform everyday behaviour. They sanction certain courses of action as 'natural' and prohibit others as 'unnatural'. This process takes place each time notions such as 'community' or 'family' are used to describe 'natural' systems, when nature is perceived to have female qualities, when humans are compared with biotic systems in terms of their 'naturally' competitive or co-operative characteristics, or when wilderness is contrasted to civilisation. Nature is consequently a powerful moral, ethical and political marker in human relationships with the environment and with one other.

- *Nature as a resource*: nature provides elements that humans can exploit. The extent to which we believe it is legitimate to use, change, discard and dispose of the materials of nature is a fundamental question for the study of environmental representations. The power to control nature is centrally important for the way in which human beings have treated nature as a resource, and has profound implications for understanding the past development and future of present-day environments.

- *Nature, an idea specific for different times and places*: just as a Himalayan farmer thinks of nature very differently than does an office worker in London's Docklands, so has nature meant different things at different times in history, even within Western society. Notions of nature pre- and post- the scientific revolution of the seventeenth century not only profoundly influenced the world of ideas and the development of technology, it also changed farming practices and attitudes towards the world's resources. We live today with the legacy of past ideas about nature as an authority and a resource. In terms of the physical landscape, we think of 'nature as a resource' for farming or industry; in the realms of the social world of 'nature as a model for human society' (e.g. are humans naturally competitive or co-operative?); and in philosophical inquiry of 'nature as an authority' (the world of ideas and the concepts of goodness, truth and beauty).

Religion, landscape and the history of nature

A paper by Lynn White (1967), published in the influential journal *Science*, provides a useful point of departure here. White evaluated human–environment relationships by critically examining the ethical basis of Western approaches to the valuation of nature. He found evidence for this in how people worked the land and created distinctive *landscapes*. For example, he noted that by the latter part of the seventh century, following obscure beginnings, certain northern peasants used an entirely new kind of plough, equipped with a vertical knife to cut the line of the furrow, a horizontal share to slice under the sod, and a mouldboard to turn it over. The friction of this plough with the soil was so great that it normally required not two but eight oxen. It attacked the land with such violence that cross ploughing was unnecessary, and fields tended to be shaped in long strips.

> In the days of the scratch-plough, fields were distributed generally in units capable of supporting a single family. Subsistence farming was the pre-supposition. But no peasant owned eight oxen: to use the new and more efficient plough, peasants pooled their oxen to form large plough-teams, originally receiving (it would appear) ploughed strips in proportion to their contribution. Thus, distribution of land was based no longer on the needs of a family but, rather, on the capacity of a power machine to till the earth. Man's relation to the soil was profoundly changed. Formerly man had been part of nature; now he was the exploiter of nature.
>
> (White, 1967, 1203)

Exercise 4.2

Examine this excerpt and briefly outline the claims made by Lynn White concerning the relationships between humans, nature, technology and landscape.

Your list should include some of the following:

- The family is the 'natural' social unit balancing production and consumption.
- Technological progress does violence to nature.
- Increased use of technology, however simple it may now seem, alienates people from their environment.
- Christianity justifies human domination of nature.

White's paper was important for five reasons:

1   It was written during the 1960s when modern society and its technological sophistication were coming under increasing criticism from environmentalists (see Chapter 1). As such, it refocused attention on a history of environmental damage dating back well before the onset of large-scale industrialisation in the West.

2   It made connections between land, landscape and ways of life, emphasising the practical relationships between humans and nature.

3   It showed ways in which changing attitudes to nature through time influenced human use and misuse of the environment.

4   White placed the origins of our environmental problems centrally within
     the sphere of Western culture and its value systems. He did so by stressing
     the relationship between environmental degradation and the grounding of
     Western civilisation in the moral and ethical values of Christianity.
5   He therefore questioned the ethical superiority of Western value systems as
     a basis for environmental action.

White used the vivid image of the new plough technology slicing into the land, as
if committing an act of violence on nature, as a metaphor for the increasingly
commercial values represented in the transition from a 'primitive' form of subsistence farming to the Medieval open field system. The open field system itself
emerged sometime during the early Middle Ages and took over much of the
landscape of lowland England as well as parts of Europe. It dominated in areas
where grain cultivation had the greatest potential and was characterised by large
open fields in which individual villagers held individual strips scattered throughout
the village lands. There was usually an agreed system of crop rotation either by
the field itself or by the *furlong*. A furlong equalled the amount of land that could
be ploughed in a day. Furlongs would be left fallow every second, third or fourth
year. The whole system was administered by a formal meeting of farmers in the
manorial court. Local by-laws regulated all aspects of agriculture: arable, pasture,
meadow, harvesting, storage, fencing, boundaries, roads and many other matters
(Donkin, 1976, 81–6; Atkins *et al.*, 1998, 140–5). Given this level of sophisticated
co-operation, it would be possible to tell a story of the development of open field
agriculture in which peace and good order are created in a previously semi-wild
landscape. In this version of events rational planning, co-operation and sharing of
resources among agricultural communities would be seen to bring great benefits.
White (1967, 1204), however, claimed that the separation from nature, due to the
rise of modern Western society based on Christian values, was the most important
factor:

> At the level of the common people this worked out in an interesting way. In
> Antiquity every tree, every spring, every stream, every hill had its own genius
> loci, its guardian spirit. These spirits were accessible to men, but were very
> unlike men; centaurs, fauns, and mermaids show their ambivalence. Before one
> cut a tree, mined a mountain, or dammed a brook, it was important to placate
> the spirit in charge of that particular situation, and to keep it placated. By
> destroying pagan animism, Christianity made it possible to exploit nature in a
> mood of indifference to the feelings of natural objects.

Though secular environmentalists, and those in the 'green movement' who
attribute current environmental problems to the legacy of Western technological
and commercial development, might concur with White's sentiments, others
criticise his arguments on both theological and historical counts:

> For Christians debate ranges around the concepts of *Dominion* and *Stewardship*. These two important concepts help define the relationship to nature within Christianity. White believes that the notion of *dominion* gives humans the power to exploit the world for their own ends in the name of the greater authority of God. Whilst Christians frequently argue that the concept of *stewardship* puts us in trust to look after God's creation and gives us a responsibility for nature and a duty of care which is quite compatible with environmentalist thinking.
>
> (Pepper, 1996, 148–55)

> Historians challenge both the timing of any changes leading to increased exploitation and the specific role played by Christianity in those changes. There are strong arguments to suggest that other technological and economic changes and other periods of history have been more important than open field farming. Christianity itself may have played a relatively small role in the lives of medieval peasants whose whole world was still dominated by nature spirits and pagan gods.
>
> (Reed, 1990, 130–2)

There are strong arguments to counter the view that Christianity was solely responsible for the development of modern economic, scientific and technological systems (Attfield, 1983), but the Church's influence extended, and indeed extends, well beyond its theological role. For more than 1,500 years, the Church exercised considerable influence in Western society through its political power, its position as landowner and as the foremost source of cultural authority. White therefore had some justification for pointing to the wider implications of religion as an influence on society's use of the environment and the making of landscape, although there is clearly more at stake in Western society's alienation from nature than simply the dominance of Christianity over a pre-existing primitive belief in nature spirits. For White, Christianity made possible a rationalist, scientific perspective on nature that was distinctively modern. Though he appeared to take a culturally *determinist* line, arguing in some way that Christianity was responsible for modern environmental problems, he also used Christianity as a metaphor representing the broader directions of Western society:

> Our daily habits of action, for example, are dominated by an implicit faith in perpetual progress that was unknown either to Greco-Roman antiquity or to the Orient. It is rooted in, and is defensible apart from, Judeo-Christian teleology ... We continue today to live, as we have lived for about 1,700 years, very largely in a context of Christian axioms.
>
> White (1967, 1203)

By suggesting that we live 'in a context of Christian axioms', White moved away from the idea of direct causality. Here religion might influence but did not control technical progress and economic development. In White's interpretation, the idea of Christianity functioned *metonymically* in a similar manner to some of the visual images discussed in Chapter 2. It acted as one part of Western civilisation that was made to stand for, or *represent*, the whole. This was paralleled in White's account by his *stereotyped* portrayal of the open field system, which actually embraced a wide variety of agricultural systems and only approached the ideal described here for relatively small periods of time in a minority of locations in Medieval Europe (Donkin, 1976). Though discussion was couched in terms of religion and the open field system, White's was more generally concerned with Western civilisation. Skilfully combined with powerful images of the plough tearing up the land, or 'primitive' spiritual life contrasted with 'modern' life, White followed Carson's lead in giving his academic paper a strong emotional appeal (see Chapter 1). By choosing landscape as the medium for bringing together his concerns about the social, economic and environmental consequences of Western 'progress', White also recognised the importance of *landscape*, alongside the concept of *nature*, in the history of human relations with the environment.

Landscape

During the Middle Ages in England, the term 'landscape' referred to the land controlled by a lord or inhabited by a particular group of people. It derived from the closely related German term *Landschaft*. Landschaft can relate to the perceived appearance of a piece of land – a landscape – but also 'a restricted piece of land' and, in some German contexts 'a region'. By the early seventeenth century, however, the influential Dutch landscape painters associated the term with the appearance of an area, more particularly to the representation of scenery. By the late nineteenth century, the contemporary definition of landscape had taken shape. Landscape was now a portion of land or territory that the eye could comprehend in a single view. Olwig (1996, 631) argued: 'It is well known that in Europe the concept of landscape and the words for it in both Romance and Germanic languages emerged around the turn of the sixteenth century to denote a painting whose primary subject matter was natural scenery.' He also argued that the northern European concept of landscape emerged much earlier than the turn of the sixteenth century and carried a range of meaning that goes far beyond natural scenery. *Landschaft* was much more than a restricted piece of land. It contained meanings of great importance to the construction of personal, political, and place identity (see also Muir, 1999, 2–24; Johnston *et al.*, 2000, 429–31).

Landscape, then, is both a physical object and a way of seeing the world. The term refers to the appearance of an area, the grouping of objects used to produce that appearance and the area itself. The term brings together meanings concerning land ownership, political and territorial control, living and working on the land, aesthetic judgements and cultural representations. Landscape is made as much by ploughing fields and building houses as by painting and writing poetry. Landscape therefore is a form of *discourse* in the sense suggested in Chapter 3, and the diverse ways of making and expressing landscapes are discursive practices that bring together the symbolic and the practical. Moreover it is apparent that we have a culturally wide and chronologically deep body of evidence on which to draw. As suggested above, the term clearly dates back to a period marking the beginning of the agricultural changes described by Lynn White, if not even earlier. Given this length of usage, the concept of 'landscape' is, therefore, extremely useful because:

- It provides a consistent frame for considering human relationships to environment over a lengthy period of time.
- It allows us to examine critically relationships between culture and nature in specific historical circumstances.
- It draws together disparate realms of experience, such as agriculture, leisure, ownership, conspicuous consumption, and artistic expression.
- It brings together the symbolic and the material, the practical and the speculative, the cultural and the natural – allowing us to focus attention on what Chapter 3 identified as the *hybrid* qualities of human activity.

In the remainder of this chapter and the two following chapters, we focus on landscape's *hybrid* status as simultaneously physical nature and cultural representation. In the process, we examine the relationships between the history of the idea of nature in Western thought and its representation in landscape, with examples drawn from the visual arts, literature, cartography and landscape design. The sections are organised chronologically.

Classical ideas of nature

The Classical period witnessed:

- the beginnings of coherent speculation about nature in Western thought;
- a sharp division between culture and nature;
- the view that humans were superior to nature;
- the idea of nature as a 'storehouse' for human exploitation;
- theologies in which many of the gods were nature based;
- sacred spaces that gave protection to nature;
- the emergence of the 'pastoral' – a valued cultivated middle ground between wilderness and the city;
- environmental damage – for example, deforestation in the Mediterranean.

In the Greek part of the Mediterranean from the sixth century BC onwards, an extraordinarily creative period of speculation and argument occurred concerning all aspects of human experience from morality and politics to religion and cosmology. As early as the fifth century BC, the physician Hippocrates of Kos made a series of observations about the connections between people and their environment in his treatise *Airs, Waters, Places*. After arguing that there was a strong link between bodily diseases and the climatic attributes of place, Hippocrates made ethnographic contrasts between Europe and Asia (in the restricted sense used by the Greeks), arguing for the superiority of temperate climates as living environments. These themes resurfaced in later environmentalist writings from the seventeenth century onwards (Arnold, 1996, 14–16). Speculation about the purpose and character of nature abounded. Some commentators refined older ideas about the divine order of things and the spiritual qualities of nature, while others sought to demolish the ancient myths and to establish a completely rationalistic view of nature (Clarke, 1993, 28). So diverse were these philosophical currents that 44 of the 66 meanings of nature listed by Arthur Lovejoy and George Boas (1935) were already current in Classical times (Coates, 1998, 23). At one extreme, the Greek philosopher Plato (428–347 BC) argued that the cosmos was a single living being, created by the gods in accordance with an eternal rational plan, in which human beings, the possessors of rational souls, had a special place (see also Chapter 7). At the other extreme, the Roman poet Lucretius (Titus Lucretius Carus, *c*. 99–55 BC) put forward a model of the universe that dispensed with the supernatural and saw the order of things as arising out of the random motion of particles.

For the Greeks, the universal god of nature was Pan – part goat, part human – who revealed his animal nature with woolly thighs and cloven feet. Derivatives of the name Pan, such as *panic* and *pandemonium*, indicate the fragile qualities of human social order in the face of nature. The name Pan itself suggested 'everything', as in *panorama*, indicating the all-embracing qualities of nature and the vitality of the animal spirit (Schama, 1995, 526–7). Like Pan, most gods and goddesses dwelt in wooded and mountainous areas (notably Mount Olympus), so places of outstanding natural beauty were invariably selected for shrines and temples. Mountaintops were set aside and embellished with a summit throne or shrine honouring the local deity. Groves were favoured sites, protected from fire, grazing, ploughing, felling, and the predations of horses and dogs. Particular trees were sacred to specific gods, for example, the oak to Zeus (Coates, 1998, 30–1). Wildlife in sacred groves was also technically off-limits to hunting, and even fishing sometimes banned. Yet although there were exceptions – such as the sacred grove at Daphne, which was ten miles in circumference – these sites were mostly relatively small in area. The vast majority of Greek (and Roman) trees and animals enjoyed no divine association and could be chopped down or killed with impunity (Hughes, 1994, 149–68; Jellicoe and Jellicoe, 1998, 117–22).

The Greek theatre influenced later landscape design. The Greek word for theatre, *theatron*, was related to words for seeing, including *thea* (sight or spectacle). The spectacle provided by early theatres was, in fact, the sight of the dramatically formed valley-and-plain landscape native to Greece. Built into hillsides, theatre was always held in the open so that a wide arc of distant landforms surrounded spectators. Indeed, this is the first example of the landscape acting literally as 'scenery'. The fact that landscape is often described as 'scenic' implies interpreting and constructing the landscape in accordance with models derived from the theatre. Moreover, many pictorial conventions for unifying a scene are adaptations dating from the Hellenistic period, when painters and sculptors imitated the Greek theatre (Crandell, 1993, 32).

According to the first-century Roman writer Vitruvius, Greek theatres had three centres of action: a central section and two projecting wings at the sides, called *parasceni* (Crandell, 1993, 35). The central section represented interior space with a building called a 'pavilion' – the modern word for a shelter in a park or garden. The wings or *coulisse*, which acted as a convention for framing the pavilion, were later transferred from the stage to painting and to landscape gardening. In eighteenth-century landscape gardens, views and open spaces were both physically juxtaposed with, and visually enclosed, by clumps of trees. This convention is still used in gardens and parks, in which clumps of trees echo one another and define vistas, thereby framing the space and giving the illusion of greater distance. Far from being natural, therefore, the principles commonly used in designing modern parks actually derive from the conventions of Classical stage design (see Jellicoe and Jellicoe, 1998, 116–37; Crandell, 1993, 36).

The pastoral

The pastoral has been one of the most important cultural influences on the understanding and management of nature since Classical times. Even though people today may not have knowledge of Classical literature, they commonly describe environments as 'Arcadian scenes', 'idyllic places' or 'pastoral land-scapes'. Among the most lastingly important texts of the Classical pastoral are Virgil's *Eclogues* and *Georgics*. Virgil was one of the first poets to write about the landscape in scenic and pictorial terms. Written between 39 and 29 BC, the *Georgics*, five poems that told of agricultural practices and farming life, were full of observations of animals and nature. The earlier written *Eclogues* (between 42 and 39 BC) comprised a series of ten short pastoral poems blending portrayal of an idealised lush, tranquil, cultivated landscape with realistic historical and geographical references to actual people and places. Their model was the bucolic *Idylls* of Theocritus written in the first half of the third century BC. Ostensibly these poems related stories of shepherds, goatherds, farmers and other rural characters and dealt with their singing competitions, amorous adventures and farming lives. Translated from Sicily to an idealised Arcadia, the rustic farming characters of Virgil's *Eclogues* inhabited a highly idealised terrain, and lived a cultured life with poetry and music. Yet, Virgil's poetry was not simple escapism. *Eclogue I* related a conversation between two farmers Meliboeus, who had been dispossessed of his land, and Tityrus, who had won reprieve from dispossession at the behest of Octavian in Rome.

> Tityrus, here you loll, your slim reed-pipe serenading
> The woodland spirit beneath a spread of sheltering beech,
> While I must leave my home place, the fields so dear to me.
> I'm driven from my home place: but you can take it easy
> In shade and teach the woods to repeat 'Fair Amaryllis'
> (Virgil, 1983, 3)

This story has direct historical reference. Virgil wrote the *Eclogues* at the time when Rome experienced the military and political upheavals that overthrew the Roman Republic and established the Roman Empire under Octavian (later the Emperor Augustus). After the defeat of Brutus and Cassius at Philippi by Anthony and Octavian in 42 BC, it was Octavian's task to find land on which to resettle and reward the returning soldiers. It is widely believed that Virgil himself was a victim of dispossession. The dispossessed Meliboeus was associated with the most precious of Roman values, *patria* – love of country. The poem told of his expulsion by an unjust military force (Patterson, 1988, 2–3). As Leo Marx (1964, 21) commented, no sooner did Virgil sketch out the ideal landscape than he referred to an alien world encroaching from without: 'Through his lines we are made aware that the immediate setting, with its tender feeling and contentment, is an oasis. Beyond the green hollow the countryside is in a state of chaos.' Deprived of his land, Tityrus faced a life of permanent exile and hardship caused by the incursion of world events into the life of rural peace:

> But the rest of us must go from here and be dispersed –
> To Scythia, bone-dry Africa, the chalky spate of the Oxus,
> Even to Britain – that place cut off at the very world's end.
> Ah, when shall I see my native land again? After long years,
> Or never? – see the turf-dressed roof of my simple cottage,
> And wondering gaze at the ears of corn that were all my kingdom?
> (Virgil, 1983, 3)

As Marx showed, *Eclogue I* clearly sketches out the geography of the pastoral ideal situated in a middle ground between the civilisation of the big city and the raw wilderness of untamed nature. To arrive at this haven, it is necessary to leave Rome, but the journey stops well short of unfarmed mountains and marshes.

> It is a place where Tityrus is spared the deprivations and anxieties associated with both the city and the wilderness. Although he is free of the repressions entailed by a complex civilization, he is not prey to the violent uncertainties of nature. His mind is cultivated and his instincts gratified. Living in an oasis of rural pleasure, he enjoys the best of both worlds – the sophisticated order of art and the simple spontaneity of nature.
>
> (Marx, 1964, 22)

The persuasive power of the pastoral rested on a fusion of realism and fantasy. Shepherds talked of real places and actual events in the same breath that they discussed the activities of deities and mythological figures, as if all these were part of the everyday world. Descriptions of an idealised land of plenty in which goats lay down to be milked and trees bent over to have their fruit picked had equal status with texts advising on the tending of bees or the construction of a plough. The result was a powerful fusion of utopianism (see Chapter 8) and realism with long-term consequences for landscape and the concept of nature. In particular:

- In the sixteenth and seventeenth centuries, the symbolic pastoral landscape was first communicated to Europeans through painting. Giorgione and Claude Lorrain, for example, explicitly depicted Virgilian scenes. Later, these artistic images took actual form in grass and trees in the English landscape garden (Crandell, 1993, 42).
- The British made comparisons between Britain and Rome, as seafaring and military powers, seeing themselves as the inheritors of Classical civilisation and possessing an Empire that was the equal of ancient Rome.
- Descriptions of farming techniques encouraged eighteenth-century English aristocratic landowners to identify themselves with Virgil's rustic characters. This encouraged experimentation and change in an age of agricultural 'improvement' and transformation.
- The ideas of Arcadia as a mythical garden landscape mapped easily on to Christian ideas of the Garden of Eden. Tityrus' dispossession echoed Adam's expulsion from Eden, while the celebration of farming as a cultured activity in the *Eclogues* and the *Georgics* reflect the Christian concept of 'stewardship' as a model for the relationship between civilised human society and nature.
- Joining Classical and Christian traditions, the idea of Arcadia as a comfortable respite from real life has been a powerful influence on European actions in the Americas from the days of early settlement, through the writings of nineteenth-century naturalists such as Henry Thoreau, to the present day. In the twentieth century the movement to create National Parks and nature

reserve areas in America, Europe and later elsewhere owes much of its inspiration to the model of Classical pastoral arcadia.

Nevertheless, this does not necessarily imply that Classical civilisations were unusually sympathetic to the ideals of nature. As Coates (1998, 35) noted, nature often appeared in Classical literature as a foil for human activity. Equally, there was more hatred of the city than love of nature (see also Chapter 7). The greatest expressions of awe and affection for wild places were connected with hunting, whilst any Greek tenderness towards wildlife was largely lost on the Romans, whose mass entertainments included the spectacle of profligate animal slaughter in the arena.

The Medieval period

In studying the Medieval period, which lasted from roughly AD 500 to 1450, we recognise:

- the dominance of Church and monarchy;
- a circular (recurring, seasonal) sense of time;
- the belief that human beings were part of a hierarchy ordained by God;
- the idea of the Earth at the centre of the heavenly system;
- God's purpose animating the world;
- all nature as religiously symbolic;
- speculation focusing on theology rather than out to the natural world;
- worldviews based on combination of Christian, Classical and other pagan thought.

Explanations for physical phenomena in Medieval Europe followed from, and were compatible with, the teachings of the Christian Church. Theology comprised by far the most important faculty in Medieval universities. The educated view of the universe was governed partly by the ideas of Classical philosophers like Aristotle (384–322 BC), but with deference to Christian theology. Classical philosophy supplied ideas about the physical qualities of the world, but Christianity provided a way of viewing this as a moral scheme ordered by God. This, in turn, supported the powerful central role played by the Christian Church in society (Pepper, 1996, 129). The Church's wealth and social importance, especially given its close alliance with the regimes of feudal kings and rulers, meant that attempts to overthrow such ideas posed a threat to the entire structure of society. One only has to consider the way in which Medieval and Renaissance science viewed the structure of the universe to see how physical order and religious ideas fitted together. This may be summarised as follows:

- The Earth, located at the centre of the universe, was solid, stationary, finite and spherical. The stars rotated around and were equidistant from Earth. They were attached to the inner surface of a rotating sphere that looked like a dome from Earth, and marked the edge of the universe.
- Outside this sphere – beyond the universe's edge – was nothing, or *non ens*. This did not mean just empty space; it meant literally that nothing could exist. One could not ask questions about this region and, by implication, one could not question either the existence of God or the power of the Church in society.
- Within the sphere of the fixed stars, the universe divided into two zones, celestial and terrestrial. Celestial objects behaved predictably; they moved in circular orbits around the Earth at constant speeds. In the terrestrial region, things moved randomly or in straight lines. Terrestrial things were born, died and decayed – they changed. This did not happen to celestial objects, for change meant imperfection. Circular motion suggested perfection. The celestial zone was one of perfection; the terrestrial zone was one of imperfection.
- All things worked towards a specific goal, meaning that there was a *final cause* behind everything (this view is called *teleology*). Since the cosmology (worldview) was a Christian one, the Christian God was the final cause. The universe was ruled by principles that helped to achieve divine purposes and God was the final cause of everything (Pepper, 1996, 127–8).

As the Medieval period progressed, Christian thought absorbed Greek philosophical ideas as theologians constructed a formal philosophy in which the natural world was given an integral place within the religious cosmology. The Aristotelian idea of the 'ladder of nature', with its hierarchical order stretching from the world of matter below to the more spiritual orders above, proved especially attractive, and generations of scholars and philosophers sought to build it into a theological worldview. According to this model, later known as the 'Great Chain of Being', humankind occupied a special median place within the chain, which ranged from matter, plants and animals below, to heavenly bodies, angels, and God above – a conception that gave everything its proper and natural place as part of an integrated and purposeful whole. This model partly overcame the sharp division between the natural and the spiritual orders, for both were viewed as elements within a continuum rather than as opposites. Consequently it gradually became possible to study the order of nature as a manifestation of God's gift to humans (Clarke, 1993, 43–4).

The twelfth century marked the key stage in the victory of the rationalistic view of nature over the mystical or supernatural one (Jolly, 1993, 224). In the early Middle Ages, the natural world bristled with messages from God, and its workings were inexplicable without recourse to an elaborate cosmology of angels, spirits and demons:

The separation of natural and supernatural made it possible to define a realm (the natural) within which human reason could seek explanation without undermining beliefs about God. God could be recognised as the primary cause of everything, but secondary causes could legitimately be sought within his creation.

(Coates, 1998, 63–4)

Jolly (1993, 226) identified a transitional period between the seventh and eleventh centuries during which Christianity progressively ousted paganism, a period paralleling that highlighted by Lynn White (see above, p. 85). This may be observed in church architecture as the vivid sculptures of mythical beasts and creatures in Saxon churches gave way to more naturalistic depiction of vegetation resembling the filigree of later Medieval tapestry in thirteenth- and fourteenth-century churches (see Figures 4.1–4.2).

Figure 4.1 *Saxon serpent capital*

In spite of the increasing dominance of the Christian perspective on nature, the pragmatism of Christian missionaries must also be recognised. Despite Biblical injunctions to destroy places of pagan worship (e.g. Deuteronomy 12.2), Pope Gregory the Great recognised the importance of absorbing the sacred places of the pre-Christian religions into places of Christian worship, both to ease the Church's

Figure 4.2 *Foliage stonework, Southwell Minster*

entry into new territories and to neutralise possible sources of competition for the people's devotion. His advice to the nascent British Church was typical:

> Do not pull down the fanes. Destroy the idols; purify the temples with holy water; set relics there; let them become temples of the true God. So the people will have no need to change their places of concourse, and where of old they were wont to sacrifice cattle to demons, thither let them continue to resort on the day of the saint to whom the church is dedicated, and slay their beasts, no longer as a sacrifice, but for a social meal in honour of Him who they now worship.
>
> (Anderson, 1971, 10)

Armed with this kind of authorisation, many of the shrewder missionaries grafted Christian theology onto pre-existing pagan nature cults.

New views of nature also reflected broader changes in the land. As the ancient forests shrank, the hostile sentiments evident in Dark Age sagas such as *Beowulf* were challenged by odes to the beauty and soothing purity of the wildwood. For example, by the time of the Domesday Survey in 1086, natural woodland perhaps covered only 15 per cent of England. The term 'forest' became equated with its legal meaning as an area of managed woodland set aside for Royal hunting, with a very different open appearance from the dense wilderness of myth and folklore (Rackham, 2001, 50; Reed, 1990, 122–5). Coates (1998, 64) argued that with the waning of the Middle Ages and the 'wildewood', the growth of the benign image of the greenwood threatened to overshadow the forest's gloomy and savage reputation. More generally, a new realism emerged across Western Europe during the fourteenth and fifteenth centuries. Originating in the great urban industrial centres of Flanders and northern Italy, this style is typified by the work of Pieter Breughel, whose paintings employed direct observation as well as tradition, memory and imagination (Clark, 1976, 55–9). However, for much of the period 'the sense of detachment and security necessary to stand back and look at the landscape appreciatively was simply nonexistent' (Crandell, 1993, 49). Clark (1976, 48) discussed this in relation to the writings of St Anselm:

> St Anselm, writing at the beginning of the twelfth century, maintained that things were harmful in proportion to the number of senses which they delighted, and therefore rated it dangerous to sit in a garden where there were roses to satisfy the senses of sight and smell, and songs and stories to please the ears. This, no doubt, expresses the strictest monastic view. The average layman would not have thought it wrong; he would simply say that nature was not enjoyable.

In cultural terms, the Middle Ages were a time of turning inward, away from the outside world, away from observation. Perhaps the most celebrated incident in the Medieval approach to nature through landscape concerned the influential Italian humanist Francesco Petrarca (Petrarch). In 1336, when in his early thirties, Petrarch climbed Mont Ventoux, one of the highest peaks in the Provence Alps. The pleasure that Petrarch took in the view, and how it inspired him to reflect on creation, were described in a letter to a friend: 'The great sweep of view spread out before me, I stood like one dazed'. However before leaving the summit, Petrarch opened his copy of St Augustine's *Confessions*, which he invariably carried. He came across the passage warning men to concentrate on their salvation instead of being seduced by scenery: 'I was abashed, and . . . I closed the book, angry with myself that I should still be admiring earthly things who might long ago have learned . . . that nothing is wonderful but the soul . . . I gazed back, and the lofty summit of the mountain seemed to me scarcely a cubit high, compared with the sublime dignity of man' (Clark, 1976, 10). Though Augustine did not teach that the earth was to be detested on account of its inferiority to God, he constantly warned against confusing the created with the creator. He believed this was the fundamental sin of paganism (see also Chapter 7). As Crandell (1993, 49) observed, the Medieval eye refused the unified point of view seen in Classical approaches to nature. In landscape depiction, this idea was demonstrated in the decorative, rather naturalistic, disposal of plants and earth forms. Flowers filled the sky like tapestry and barren rocks were piled up to act as stages for events in a linear story. The Medieval tendency was to picture nature in terms of these identifiable objects rather than in terms of views or scenes. Not surprisingly, the period supplied relatively little evidence for the representation of nature through landscape until as late as the fourteenth century, even though the human impact on the land itself gained increasing momentum (Jellicoe and Jellicoe, 1998, 138–9; Reed, 1990, 116–63).

Figure 4.3 **Les Très Riches Heures**
(1409–15) by the Limbourg brothers

Figure 4.3 depicts one of a series of twelve illustrations in illuminated manuscript by the Limbourg brothers from Nimwegen (now in Belgium). *Les Très Riches Heures* is a classic example of a Medieval book of hours. This was a collection of the text for each liturgical hour of the day and often included other supplementary texts, including calendars, prayers, psalms, and masses. The twelve illustrations were from the calendar section, and formed one of the most important early sources in the history of landscape painting. They were painted for the Duke de Berry, one of the highest nobles of fifteenth-century France and reputedly one of the Medieval world's greatest art connoisseurs. This illustration was for August and showed an aristocratic hunting party dressed in fine clothes and carrying falcons. In the background peasants attended to the harvest and swam in the river. Behind them was the Chateau d'Etampes. Things to note include:

- The artists set scenes of everyday life for both peasants and aristocracy within a vertically organised frame with a representation of the heavens at the top of the picture. The Chateau, as residence of the aristocrats, was located towards the top of the image, emphasising the relationship between aristocracy, state and church. This reproduced the Medieval worldview of a vertical, stable and hierarchically organised society, in which church and aristocracy mediated between heaven and earth.
- The scene portrayed both work and leisure within an orderly agricultural landscape, watched over by the feudal ruler. The artists made pictorial connections between an idealised, safe and bounteous garden landscape, the good Christian stewardship of the feudal magnate, and Christian ideas of a paradise garden.
- Each of the twelve images represented some form of farming or practical everyday activity, more than half represent work in the fields. Clark (1976, 22) believed these were the best representations of agricultural life in the Middle Ages, indicating keen observation of everyday life on the part of the artists and placing these works some way towards the developing portrayal of landscape factually that was to become important in the Renaissance. He argued that factual observation set within a Medieval cosmology places these works 'half way between symbol and fact'.

With the collapse of the Western Roman Empire and the beginning of the Dark Ages (c. sixth century), garden art virtually disappeared (Thacker, 1994, 16). During the Middle Ages, the garden again became prominent in Western life. Twelfth-century gardens were square, enclosed and, like all Medieval art, symbolic (Taylor, 1983, 33–40). In religious houses a walled garden was the symbolic environment of the Virgin Mary (Harvey, 1981, 27). Enclosed, inwardly oriented and square gardens dramatically paralleled the iconographic developments occurring in pictures during this period. Gina Crandell (1993, 58) maintained that the atmospheric distances and city vistas of Roman painting and the outward, spatial connections to the landscape of Greek site planning are, in both the paintings and gardens of the Middle Ages, turned inward, enclosed, and flattened. Medieval gardens were set apart from the world that lay just outside, choosing not to look at distant prospects of landscape.

In common with paintings and other forms of representation, maps yield a great deal of information concerning environmental actions, values and their histories. The map shown in Figure 4.4 is typical of the fusion of Christian, Classical and mythical knowledge that formed the Medieval worldview. It is possible that this is a miniature copy of the map that the English King Henry III commissioned as a mural for the walls of Westminster Palace in 1236 (Whitfield, 1994, 18). Maps like these are known either by their Latin name *mappae mundi* ('maps of the world')

Figure 4.4 *Psalter Map*, c. 1250.
Add. 15,500.8, courtesy of the British Library

or as 'T-in-O' maps, after the configuration of the continents shown – with the Mediterranean forming the stem of the letter 'T', the rivers Don and Nile the horizontal bar, and the land areas of Europe, Asia and Africa assuming the 'O' shape (Brown, 1977, 100–6). They primarily related a narrative with a historical theme rather than showed location, in a manner similar to the stained-glass windows of the great European cathedrals. The stories usually related either Christ's dominion over the face of the Earth or illustrated certain Old Testament tales (Dorling and Fairbairn, 1997, 13). This map illustrated motifs that became characteristic of the elaborated *mappae mundi*:

- It was one of the first maps to place Jerusalem firmly at the centre of the world and to symbolise Christ's power as overseer of the world.
- It depicted Biblical events such as Noah's Ark, the crossing of the Red Sea, and the walls imprisoning Gog and Magog. It was also among the first to display what were known as the 'monstrous races' in Africa (see also Chapter 6).
- It showed the figure of Christ dominating the world and symbolically holding in his hand a small T-O globe. This developed the idea of Christ the Pantocrater, 'ruler of the whole', that had emerged in Byzantine art.

- Its inspiration was Biblical rather than geographical. Given that the twelfth-century Crusades brought new geographical awareness into European society, as thousands of west Europeans crossed and re-crossed half of the known world, it was surprising that maps like these remained static in form and content (see Whitfield, 1994, 18).

Renaissance and Early Modern periods

In looking at the Renaissance and Early Modern periods, approximately dealing with the years from 1450 to 1650, we note:

- the beginnings of modern world system of international trade;
- the rediscovery of Classical science;
- the beginnings of science based on practical observation and mathematical calculation;
- empirical knowledge of nature in plant collecting and exploration driven by trade and commerce;
- practical scientific methods in religious and mystical confusion of Classical and Christian sources – the alchemists;
- challenges to dominant Christian beliefs about Earth – Heliocentrism, Copernicus, Galileo;
- the development of landscape as a highly sophisticated means of representing, ordering and controlling nature.

The Renaissance matched the extraordinary period of creativity of Classical times, with profound developments in the arts, literature and culture. Though this eventually gave rise to an 'enlightened' humanism (a system of thought in which human interests, values and dignities assume prime importance), Renaissance thought delved into mystical, even occult, ideas, while remaining steeped in Christian theological traditions (Glacken, 1967, 462–9). Following Medieval developments of ancient teachings, Renaissance thinkers tended to see the natural world not as a collection of material objects, but as a sort of text comprising a whole interlocking set of signs and symbols, like a book to be read, and in which could be discerned the mind of and intentions of the Creator. The world was seen as a manifestation of Spirit, 'the vital principle', the meaning of which was encoded in the beauty and harmony of nature. It was effectively a kind of magical universe. Its signs could not only be read but, in so far as they manifested hidden powers and influences in nature, could be exploited for human purposes (Clarke, 1993, 63).

Figure 4.5, by the leading Italian landscape painter Piero di Cosimo (1462–c. 1521), shows one of the earliest paintings to take nature rather than people as its main subject. It depicted a nature that was highly symbolic, since the picture told an allegorical story (Clark, 1976, 77). As such, it is a complex painting, for instance:

- If the subject of the painting was fire and panic, why did many of the birds and some of the animals in the foreground seem so unconcerned?
- Why was there a strange area of baking embers on the left, far away from the fire?
- How can we explain the bizarre details of a deer and a pig with apparently human faces?

Figure 4.5 The Forest Fire *(c. 1505) by Piero di Cosimo (1462–c. 1521). Courtesy of Ashmolean Museum, Oxford*

Cosimo trained in the Florentine tradition and enjoyed a reputation as an artist of great originality and fantasy – especially in his paintings of scenes from allegories and Classical myths (Fermor, 1993, 7–12). Figure 4.5 shows the third in a series of five paintings based on an interpretation of Book V of Lucretius' *De Rerum Natura*, an important book of Classical science. Taken literally, the painting showed the destruction in nature created by fire when unleashed by humans. It represented the point in *De Rerum Natura* where Lucretius attributed the discovery of fire to the accidental rubbing together of branches. Among the animals escaping from the flames are two with human faces, which exemplify a theory attributed to the philosopher Democritus that, in the early days of creation, nature had not decided how the various features of the animal world should be allotted (Clark, 1976, 77). However, a more complex story unfolds when considering the symbolic qualities of the picture more fully. *The Forest Fire* comprised three sections:

- The first section to the left of the painting seems to symbolise the earliest phase before man was a part of nature. In this section fire is absent, and the beasts are looking towards the centre of the picture where the fire is burning. The deer in this part of the painting evoke longevity, contrasting the deer in other parts of the painting that are fleeing from the fire. Towards the middle of the painting, animals begin to take on human qualities.

- The centre of the painting is engulfed in flames and the animals seem to be in the process of flight. In this part of the picture fire may be seen to symbolise rebirth, a creative force for vegetation and also for purification. A lion is looking inward to the wilderness burning in the centre of the painting. The lion may be connected into the symbolic properties of fire such as the divine, cleansing and the destruction of evil.
- The right side of the picture suggests an allegory for the triumph of civilisation. The female lion suggests resurrection, symbolised as an animal that eats the sun at the end of the day. It stands next to the heron which symbolises the dawn. The man in this part of the picture carries a yoke, a symbol for farming and the domestication of animals, the tree here has been cut with an axe perhaps suggesting that humans have learned to harness nature and possibly fire, whilst in the background there are images of a house and a family. The humans are dressed and have footwear, suggesting that they live in a state of relatively advanced civilisation.

The theme of sound forms a unifying theme to the picture, the ox bellows, the bear and lions are open mouthed and the birds in flight suggest noise and movement; even the herdsman is open mouthed shouting to the gesticulating figures in the distance. Whistler and Bomford (1999, 18) believed that this referred to the part of *De Rerum Natura* that discussed how language evolved. The picture followed quite closely Lucretius' description of how birds and animals communicate and showed the role of fire as a catalyst for the development of human communication and, by implication, social relationships and responsibilities.

A second description of a forest fire occurred a little later in Lucretius' poem, in which early humans learned the nature of metals. This moved closer to the more advanced stage of civilisation represented in *The Forest Fire* where humans have already mastered the use of metals:

> When a fire had burnt down vast forests with its heat on mighty mountains, either when heaven's lightning was hurled upon it, or because waging a forest-war with one another men had carried fire amongst the foe to rouse panic, or else because allured by the richness of the land they desired to clear the fat fields and make the countryside into pasture . . . for whatever cause the flaming heat had baked the earth with fire, the streams of silver and gold, and likewise of copper and lead, gathered together and trickled from the boiling veins into hollow places in the ground. And when they saw them afterwards hardened and shining on the ground . . . then it came home to them that these metals might be melted by heat.
>
> (cited in Whistler and Bomford, 1999, 20–1)

The idea of a raging fire, and of veins of metal erupting from beneath the earth and hardening into stones, perhaps lay behind the otherwise mysterious left-hand area of the painting. Here, some distance from the actual fire, the earth was red-hot, while the charred tree-trunks, the burnt grass and the groups of stones across the foreground remained after the earth has cooled. The artist might well have wanted to remind the viewer that fire has acted as a catalyst for the progress of civilisation.

Taken as a whole, therefore, *The Forest Fire*:

● comprised a rich symbolic text combining Classical scientific and religious symbolic imagery;
● depicted the natural element of fire as a destructive and creative force;
● interpreted the rise of human civilisation as a triumph of technology and ingenuity;
● suggested humans have responsibility to look after the world in their charge.

The ideas of a cosmic harmony that the previous section showed as being encapsulated in the notion of the 'Great Chain of Being' and in the *organic metaphor*, which saw the world as a huge animal with feelings, were fundamental for relationships between humans and the environment. The idea of the Great Chain of Being placed people and nature in an intimate relationship. Interconnections were important; to eliminate one link, one creature, or one part of the inanimate matter would damage the harmony of the universe and disturb the ordering of the world as God had decreed it. The organic metaphor produced a concept of nature in which all elements were part of a gigantic body with anthropomorphic characteristics. Mountains were 'brows', 'shoulders' and 'feet', rivers were 'heads', 'gorges' or 'mouths'; water formed part of a circulatory system like that in the human body (Pepper, 1996, 134).

Both concepts resulted in some degree of humility and respect towards nature. There are some similarities here with the present-day notion of an ecosystem and the ideas of the Earth as a living system made in James Lovelock's Gaia hypothesis (Lovelock, 1988). This had important implications for the exploitation of the world's resources. The belief that the Earth produced minerals and metals within her reproductive system like a living organism, persisted from Greek and Roman writers until well into the eighteenth century. Several ancient writers warned against mining the depths of Mother Earth and Pliny in his *Natural History* (*c.* AD 78) had argued that earthquakes are an expression of anger at the violation of the Earth, for example: 'We trace out all the veins of the Earth and yet are astonished that it should occasionally cleave asunder or tremble: as though these signs could be any other than expressions of the indignation felt by our sacred parent!' (cited in Gold, 1984, 14). Such restraints against the exploitation of resources remained active until the Renaissance, but lessened as commercial mining gathered momentum during the fifteenth century. Changing attitudes towards the exploitation of resources may be witnessed in the fortunes of the idea of 'Mother Nature'.

Mother Nature

Mick Gold (1984, 15) related an allegory published in Germany in 1495, in which the conflict between respect for the Earth and the commercial interests of mining was dramatised in the form of a vision. A hermit fell asleep and dreamed that he witnessed a confrontation between a miner and Mother Earth – 'noble and freeborn, clad in a green robe, who walked like a person rather mature in years'. Her clothing was torn and her body pierced. Several gods who accused the miner of murder accompanied her. In his defence, the miner argued that Earth 'who takes the name of mother and proclaimed her love for mankind' in reality concealed metals in her inwards parts in such a way that she fulfilled the role of a step-mother rather than a true parent. Thus, Gold concluded that in these early stages of capitalism, nature's image was changing from something to be respected (the mother) into a source of wealth that needed to be forced into revealing things (the selfish step-parent).

The idea of nature as a female, protecting the environment in her guise as Mother Earth, was reversed in the seventeenth century to justify the exploitation of natural resources. Carolyn Merchant (1995) drew attention to the work of Francis Bacon (1561–1626), the pioneer of scientific research methods and scientific reasoning, Lord Chancellor of England under King James I, and the author of *Daemonologie* (1597), a book on witchcraft. In 1603, King James passed a law condemning all practitioners of witchcraft to death. Merchant argued that Bacon proposed an experimental methodology for investigating nature, using language that was starkly sexual in its metaphors and suggestive of a witch-hunter in its techniques. Bacon and his contemporaries combined ideas of women as 'mere animals' and agents of the devil with those characterising them as wild and untameable, withholding 'favours' and 'secrets' from men as part of some spiteful and flirtatious game. This transformed the concept of Mother Nature from one of reverence and respect to one of exposure, discipline and correction. Metaphors of the Female became a tool in adapting scientific knowledge and method to a new form of human power over nature and were instrumental in the transformation of the Earth as a nurturing mother and womb of life into a source of secrets to be extracted for economic advance (Thomas, 1983, 43; Merchant, 1995, 80). These ideas assisted exploitation of the natural environment. Nature was a woman's womb harbouring secrets that men could wrest from her grasp using technology supposedly for the improvement of human conditions. Gold (1984, 16) remarked on the striking contrast between Bacon's attitude towards nature and the image of the Earth as an elderly lady who had been assaulted (as reported in the hermit's dream of 120 years earlier). In it one saw the characterisation of nature as a woman changing to allow research and exploitation to take place. Bacon even described physical matter as a 'wanton harlot', nature must be 'bound into service', made a slave, 'put in constraint' and 'moulded by the mechanical arts'. Mercantile capitalism viewed nature as a resource to be exploited, as can be seen from the example of mining, whilst scientific research produced a mathematical model of the universe, which superseded the organic analogies of the animate Earth.

The factual landscape

Landscape painting first emerged in the two most economically advanced, densely populated and highly urbanised areas of Europe: Italy and the Low Countries (Whyte, 2002, 56). Sixteenth-century artists and their patrons lived in an urban civilisation that had learned how to exert some control over natural forces. Crandell (1993, 83) argued that, as a consequence of this, they could look at the landscape with a certain detachment not felt before. Storms, forests, floods, mountains and fire became subjects both distant and exciting. The security that artists sensed in their urban environment, combined with their determination to explore new avenues of landscape depiction, led them to paint landscapes that were both pleasurable and horrifying. These paintings were intended to incite emotion rather than relate information, and they were infused with the idea of wilderness, of that which is bewildering and uncontrollable. This pictorialisation of terrifying natural forces in order to generate excitement was clear evidence of an increasing confidence in the ability to control such forces. The idea of solitude in a natural setting became a powerful attraction for writers and painters at the point where the citizen began to sense that there was a profound division between the civilised and the natural. As Andrews (1999, 31) observed: 'Much of the literature of Augustan Rome and Elizabethan England, and many paintings of the early Italian and Northern Renaissance, celebrate the fugitive from civilization, the figure who recovers a kind of spiritual integrity and wisdom by immersion in the natural setting.' In the early fifteenth century artists such as Jan van Eyck began painting detailed and accurate landscape backgrounds in paintings whose main subjects were religious and historical. Artists used landscape backgrounds to show their technical and imaginative skills on pictures that were still primarily about religious subjects. In the late fifteenth and early sixteenth centuries, landscape assumed a more independent role, particularly as new artistic movements appeared in Italy and the Low Countries.

Italy

Italian artists combined Classical and allegorical subjects with knowledge of geometry, cartography and formally composed single-point perspective. The rediscovery and reworking of geometry was particularly significant, with the development of visual perspective closely linked to the practices of map making, land surveying, cartography and navigation. Following the development of perspective and the panorama in central Italy, progress in landscape painting during the sixteenth century moved northward, especially to the area around Venice – where the term landscape (*paese*) was first applied to an individual painting (Cosgrove and Daniels, 1988, 254). The re-evaluation of Euclidian geometry, or more specifically its application to three-dimensional space through single-point perspective theory and technique, was fundamental to the design and representation of landscape. The architect Alberti, for example, developed a method that he had worked out experimentally for constructing a visual triangle. This allowed the painter to determine the shape and measurement of a gridded square placed on the ground when viewed along the horizontal axis and to reproduce in pictorial form its appearance to the eye. This technique, the vanishing point, distance point and intersecting plane are now familiar parts of both technical drawing and fine art.

Figure 4.6 illustrates these principles by reference to *Hunt in the Forest* by Paolo Uccello (1397–1475). This painting depicted the leisured aristocratic pursuit of hunting. In this sense, it clearly looks back to similar Medieval representations, while simultaneously reflecting the interests and lifestyles of wealthy patrons in Renaissance Italy. Yet hunting was frequently used as an allegory of 'Love'. Uccello painted it for the private rooms of Lorenzo de'Medici in the Medici Palace in Florence. As a leading poet and philosopher, Lorenzo was fully aware of its allegorical significance. Indeed his own sonnets reflect his quest for love and this was often to be found in harmony with nature (Ashmolean Museum, 1986).

Viewed technically, the painting revealed Uccello's contribution to the development of perspective and its establishment as a major component in Renaissance style. It was based on a complex multipoint system of perspective:

- The branches lying in the foreground established a central vanishing point P on the horizon.
- A series of secondary vanishing points across the image diminish the overwhelming impact of depth that would have resulted from a single focal point. By linking secondary vanishing points to the main one, Uccello succeeded in conveying optically a feeling of increased remoteness towards the edges of the panel.
- The whole panel could be divided into three large perspective boxes each dependent upon the trees in the foreground. In each of the two outer boxes the placing of the more distant trees was carefully calculated. By such means Uccello encouraged the eye not only to move towards the centre of the panel

but also to explore its outer limits. This was exemplified by the zigzag line created by the recession of trees, and by the wide spacing of the figures in the foreground, allowing the eye to be attracted towards the smaller distant forms contained in the flanking boxes.

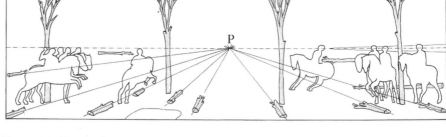

Figure 4.6 Hunt in the Forest *(c. 1465–70) by Paolo Uccello (1397–1475). Courtesy of Ashmolean Museum, Oxford*

The *Hunt* showed that Uccello's principal aim 'was not simply the application of extremely complex theories of perspective, but rather to inquire into the nature and validity of this new method of depicting the real world' (Ashmolean Museum, 1986). In this picture, the forest as wilderness is transformed into an ordered landscape by applying the scientific principles of geometrical perspective. Echoing both the codes of aristocratic hunting and the practice of romantic courtship in Renaissance Italy, the spectator could enjoy the chase while recognising the security of established, rule-governed behaviour.

These changing artistic conventions, therefore, hinted at the wider changes taking place in society. In a related way, the growing popularity of landscape painting in the sixteenth century, where an expansive view of countryside began to predominate over any narrative elements, can be related to the contemporary growth of interest in villa life in Italy. In the Venice region, for instance, the state was very closely involved in draining the 'Terraferma' – an area of low-lying, poorly drained and marshy land on the mainland. Although motivated primarily by the need to grow grain, reclamation encouraged Venetian nobles to purchase estates outside the city and have villas built on them. Andrews (1999) argued that views of the countryside from the terrace or loggia, or through the frame of a window, encouraged the pictorialising of landscape. The Villa Rotunda at Vicenza, built in the early 1550s to the design of Andrea Palladio (1508–80), differed from many villas built earlier in that it had no connection with farming (Ackerman, 1990, 106; also Cosgrove, 1993). It was built purely for the pleasure of a retired official of the papal court, Monsignor Paolo Almerico. The priorities for its setting were largely aesthetic. Palladio in his *Quattro Libri* described these:

> The site is one of the most pleasant and delightful that one could find because it is at the top of a hill with an easy ascent and is bathed on one side by the Bacchiglione, a navigable river, and on the other is surrounded by the most agreeable hills which give the aspect of a great theatre: and all are cultivated and abound in most excellent fruits and the best vines. Thus, because it enjoys beautiful views on every side, some of which are limited, others more distant, and still others that reach the horizon, loggias have been made on all four sides.
>
> (quoted in Andrews, 1999, 56)

The new villas occupied locations with panoramas of the surrounding country-side. Their main reception rooms, often on the first floor, took advantage of the views. Most were the centres of functioning agricultural estates, but villas on the Venetian Terraferma were also summer retreats for the urban nobility who patronised scholars, writers and poets. Estates were seen as refuges from the pressures of urban life, and from the dangers of the ever-present threat of plague. The idealisation of the countryside associated with these and other pressures accompanied by the renewed interest in Classical learning resulted in a resurgence of the Classical pastoral (see above, p. 90). With such close links between literature, landscape art, gardening and architecture, it was not surprising that villa design became a central element in thinking about nature in Renaissance Italy (Whyte, 2002, 58).

The Low Countries

By contrast with Italy, painters in The Netherlands, Flanders and the Low Countries concerned themselves more directly with depicting rural and urban scenes from observation and the portrayal of everyday life, including scenes of the peasantry and agricultural labourers. As with Italy, this development needs to be viewed against the political and intellectual climate of the times. After the Counter-Reformation resulted in a victory for Protestant Christianity over Catholicism, the renewed freedom to indulge in science resulted in what has been called the Age of Observation. This included investigations of nature by microscope and botanical classification, although this science was more empirical than theoretical. Seventeenth-century Dutch painting reflected this, producing naturalistic perspective without overt mathematical theorisation. As in northern Italy, money from the cities financed the development of commercialised agriculture on reclaimed land, in this instance behind the coastal dunes and along the Scheldt estuary. When Flanders was devastated by war, many painters moved north into the Dutch Seven United Provinces, which had declared independence from Spain in 1579. A Protestant (Calvinist) religious culture and a middle class of merchants and ship owners provided a ready market there for secular art. As Whyte (2002, 60) noted, it was no coincidence that the Dutch golden age of maritime expansion was also an age of flourishing art.

Figure 4.7 shows the World Map by Pieter van den Keere. Drawn in 1611, it constitutes an archetypal world image from the age of maritime war, trade and science. Three features are particularly notable:

1   The maker drew from all four known continents to represent 14 cities, 18 pairs of representative figures, and the rulers of seven nations, including China, Turkey and the Holy Roman Empire. No doubt these portraits were stylised, but it was the geographical scope that is important. These were monarchs that controlled the world's destiny, and these were the cities that they ruled. Inside this framework were the six symbolic female figures of the continents, with Europe at the centre receiving the fruits of art, war and trade.
2   In the Pacific Ocean, the largest allegorical tableau alluded to the truce negotiated between Spain and The Netherlands in 1609: the figure of Peace is crowned by Victory, while Justice holds War in chains.
3   The map was effectively a paper theatre of the world, enabling the Amsterdam merchant to contemplate at leisure the kings, the cities and the peoples of the Earth. Like the Medieval *mappae mundi*, it was a cultural icon, an image of the world interpreted according to the knowledge and values of its time (Whitfield, 1994, 80).

Dutch patrons required paintings of an unadorned, realistic rural world, flat landscapes with large areas of sky. Paintings of this era were often fully recognisable, or only slightly altered, landscapes. Between 1610 and 1679 the number of landscape paintings sold rose from 27 per cent to 41 per cent of all paintings. Jensen Adams (1994, 40) argued that, historically, the Dutch maintained a unique and tangible relationship with their land, quoting a popular Dutch saying: 'God

Figure 4.7 *World Map by Pieter van den Keere (1611)*

created the world, but the Dutch created Holland'. From the late sixteenth century onwards the United Provinces undertook the most extensive land reclamation project yet seen. Between 1590 and 1664 they reclaimed more than 110,000 hectares of land from the sea and inland lakes by means of a complex system of dykes and drainage. The drainage projects involved the joint efforts of groups of citizens and the involvement of municipal governments (Jensen Adams, 1994, 57). In this way, drainage, land reclamation and improvement came to symbolise civic pride and a citizenry independent of religious and aristocratic control.

Figure 4.8 shows a painting of the village church of Middelharnis in the province of South Holland by Meindert Hobbema (1638–1709). The view is remarkably accurate and has changed little since the seventeenth century. Before 1660, Hobbema had been the pupil of Jacob van Ruisdael but by 1689, from which this painting dates, Hobbema was no longer a professional artist as he had obtained a well-paid job with the wine-importers' association of Amsterdam in 1668. From that point, he painted only occasionally, with this picture one of a handful of pictures from this later period. The picture was probably painted at the behest of the town council, which had the avenue of trees planted in 1664.

Figure 4.8 The Avenue, Middelharnis *(1689) by Meindert Hobbema (1638–1709).*
Courtesy of National Gallery, London

Exercise 4.3

Why do you think the town council would want a painting of a tree-lined road?

How does the landscape differ on either side of the road?

Things to note about this picture include:

- Unlike many similar landscape images this avenue did not lead to a stately home; even the local church is displaced to the left of the picture. The avenue now becomes the main theme of a picture. Hobbema's rows of trees are an independent formation.
- The Avenue appeared to divide wild nature, represented by the trees and undergrowth on the left, from the enclosed nature represented by the regular planning and farm buildings on the right, suggesting a continuing process of landscape improvement and reclamation.

- Together the road and the trees (the subject of the painting) represented civic pride and effort relating to land reclamation, environmental improvement and economic enterprise. As Warnke (1994, 16) commented: 'Proudly erect and freed from courtly subservience, the trees line up as witnesses to a cultural achievement of the Dutch republic'.

Such pictures were a visual appropriation of the countryside (in which there was always danger of flooding) by an increasingly prosperous and self-confident urban bourgeoisie, whose roots in the land were often close and who were actively investing capital in rural landscape change. In a land with a high proportion of immigrants and no monarch to symbolise national identity, Jensen Adams (1994, 65) argued that the Dutch turned to their land in the creation of a largely secular communal identity.

Exercise 4.4

List the reasons why landscape was central to Dutch ideas of national identity.

Your list might well include:

- Reclamation of the land from the sea represented the physical human effort, practical and technical skill and ingenuity of the Dutch people.
- For a trading nation built around commercial values, land constituted real wealth and is therefore a source of pride.
- Protestant religion values a work ethic as the basis of its moral code and thus evidence of hard work was also a celebration of spiritual values.
- For a new nation trying to rid themselves of the symbols of the past, which represent the previous oppressive regime, land formed a focus for value. It was also a symbolic rallying point, which is historically neutral but politically positively charged given the extent to which the Dutch as a nation see themselves as born under a constant environmental threat from the sea.
- Together, these added up to a developing myth of national identity. In this myth, the Dutch were a self-made people with a purpose and a mission to reclaim land from the sea. (There are parallels here with the idea of 'manifest destiny' that inspired European colonisation of North America.)

Conclusion

This chapter began by arguing that ideas of *nature* and *landscape* were central to the history of Western environmental values. It has shown how ideas of landscape and nature developed during Classical times from the sophisticated literate culture based in a mythical and pagan spirituality. During the Medieval period, the dominating influence of Christianity resulted in a philosophically and morally inward looking approach to nature. The outside world was shunned as a source of inspiration at the same time that nature still remained a considerable physical threat. Increasing confidence in human ability to control nature during the Renaissance coupled with the rise of science based on the rediscovery of Classical learning helped produce highly complex landscapes as physical arrangements of the land and in painting and literature. The practical representation of landscape in cartography and surveying was fundamental to both landscape design and its artistic representation.

As some of the examples in this chapter have shown, the history of environmental exploitation is both conceptually and practically much more complex than Lynn White seemed to suggest. White, however, was right to direct our attention towards landscape as a medium for understanding environmental values and most importantly to recognise the significance of concepts of *dominion* and *stewardship* as key to debates concerning rights and responsibilities, the limits and possibilities for environmental exploitation. This debate is still central to ethical and political debates concerning the environment (see for example Baird Callicot, 1994).

Exercise 4.5

Briefly review the examples used in this chapter, noting the extent to which each justifies either dominion or stewardship as the basis for environmental action.

As will be shown in the next chapter, many of the environmental values developed during the period considered in this chapter surface again as they are changed and developed during later periods of European history. Nevertheless, it is worth highlighting three emerging themes central to environmental representation and fundamental to the later development of environmental values and actions:

1   *The pastoral*: links Classical and Christian traditions to form a very powerful ideological formation combining the Arcadian rural idyll with the Christian Garden of Eden. Among many other things, this has been important for:

- The eighteenth-century English landscape garden.
- European colonisation in North America.
- In the twentieth century, the idea of a middle ground between City and Country was important for the development of suburbia, the Garden City Movement, National Parks, Green Belt planning legislation and the history of public open space.

2 *Animate nature*: links ideas of nature as an expression of spirituality and religious values with a model of the world as an interdependent whole, a multiplicity of parts which make up one single living organism. These ideas have been important for approaches to the environment since the eighteenth century, since they make connection between humans and nature in emotional terms. These ideas are also relevant to the development of ecological science, which studies the interdependences between different parts of the Earth as a 'living system'. For example:

- Since the beginning of the twentieth century the rediscovery of spiritual, emotional and functional ties between humans and nature has been important for the environmental movement and for a wide range of outdoor pursuits from gardening and camping to hiking and mountaineering in which closeness to nature provides us with spiritual and physical renewal.

3 *Landscape and identity*: links land as a way of life, a representation of the everyday routines of ordinary people, and land as property, a source of material wealth and property, to land as a repository of history and memories. This has been important, for example, in numerous myths of national identity:

- England as a garden landscape of pretty villages.
- The winning of the 'Wild West' in the USA.
- Since the nineteenth century, landscape as a source of identity has been an important justification for landscape preservation and conservation. In addition landscape as a marker for status and lifestyle aspirations has been important in film, television and advertising.

Further reading

For the relationship between landscape form, history and design with an international perspective, see:

Peter Atkins, Ian Simmons and Brian Roberts (1998) *People, Land and Time: an introduction to the relations between landscape, culture and environment*, London: Arnold.

Geoffrey Jellicoe and Susan Jellicoe (1998) *The Landscape of Man: shaping the environment from prehistory to the present day*, London: Thames and Hudson.

Yi-Fu Tuan (1974) *Topophilia: a study of environmental perception, attitudes and values*, Englewood Cliffs, NJ: Prentice Hall.

On the relationship between landscape meanings, culture and art in Europe, see:

Malcolm Andrews (1999) *Landscape and Western Art*, Oxford: Oxford University Press.
Gina Crandell (1993) *Nature Pictorialized: 'the view' in landscape history*, Baltimore, MD: Johns Hopkins University Press.
Simon Schama (1995) *Landscape and Memory*, London: HarperCollins.

On the relationship between nature and the history of Western thought, see:

Peter Coates (1998) *Nature: Western attitudes since ancient times*, Cambridge: Polity Press.
Clarence Glacken (1967) *Traces on the Rhodian Shore: nature and culture in Western thought from ancient times to the end of the eighteenth century*, Berkeley: University of California Press.
Mick Gold (1984) 'A history of nature', in Doreen Massey and John Allen, eds, *Geography Matters! A reader*, Cambridge: Cambridge University Press.
Phil Macnaghten and John Urry (1998) *Contested Natures*, London: Sage.
Keith Thomas (1984) *Man and the Natural World*, Harmondsworth: Penguin.

On the implications of the pastoral for ideas of nature in the USA and the UK, see:

Leo Marx (1964) *The Machine in the Garden: technology and the pastoral idea in America*, New York: Oxford University Press.
Kenneth R. Olwig (2002) *Landscape, Nature, and the Body Politic: from Britain's renaissance to America's new world*, Madison: University of Wisconsin Press.
Raymond Williams (1973) *The Country and the City*, London: Chatto and Windus.

5 Enlightenment and Romanticism

This chapter:

- follows on from the discussion of Classical, Medieval and Renaissance environmental values and representations in Chapter 4;
- traces the development of Western environmental values associated with Enlightenment and Romanticism to nature and landscape;
- leads on to the discussion of imperialist approaches in Chapter 6.

Enlightenment, 1650–1800

The typical ideas from the Enlightenment highlighted in the first part of this chapter, include:

- Nature is like a machine.
- Humans are separate from nature because they have logic and understanding.
- The universe obeys mathematical laws.
- The decentring of humans in a heliocentric (sun-centred) solar system.
- Science becomes increasingly free from values of religion.
- The increasing importance of 'rational calculation' informing human attitudes to nature.
- The growing importance of secular linear time.
- Increasing focus on ideas of progress, improvement and control concerning human interventions in nature.

The idea that nature was a vast machine rather than a giant organism inspired the scientific revolution, and, as a potent metaphor, runs through the whole of modern thought and culture. During the sixteenth century a new and dramatically different model of the universe gradually emerged. In 1543, Copernicus argued that the Earth rotated daily on its axis, as well as orbited annually around the sun. He thereby destroyed the image of a cosmos that had symbolised humanity's central place of humanity within the Great Chain of Being. However, the main task of building the new model fell to Kepler, Galileo and Newton:

> Between them, over a period of about a century, they showed that the processes of the natural world, whether in the heavens or on earth, could be understood without reference to soul, or to purpose, but could simply be understood in terms of material particles moving in infinite space in accordance with strict, mathematically precise, universal laws.
>
> (Clarke, 1993, 81)

Isaac Newton laid out the culminating synthesis in *Principia Mathematica*, first published in 1687. The French philosopher René Descartes articulated clearly the analogy between the workings of nature and the workings of a giant clockwork machine. To Descartes:

- The mind and body represented two completely distinct kinds of substance, and that mind (defined as thought or consciousness) was therefore completely excluded from the natural world. The conceptual separation of mind (consciousness) and matter (the physical world) is often referred to as 'Cartesian dualism'.
- Living things were automata, the difference between humans and animals being that the former have a conscious mind, whereas the latter did not. Coates (1998, 76) argued that the implications of 'Cartesian dualism' were gravest where animals were concerned opening the way for experimentation without anaesthetic for anatomical investigation (vivisection).

In the eighteenth century, new ideas of the perfectibility of nature flourished. Refined taste in landscape, as in architecture, was informed by a mechanistic conception of nature as a well-regulated and predictable system (see Figure 5.1). The idea of proportionality and symmetry was embodied in the Palladian style (see Chapter 4), and exemplified by the palace and grounds of Versailles. The formal gardens of eighteenth-century continental Europe signified the triumph of culture over a self-willed natural world (Coates, 1998, 117). Straight avenues radiated from a central axis formed by the house itself. Pumps forced water uphill and into fountains. The outdoor art of topiary distorted trees and bushes into every conceivable shape. Geometrically shaped garden terraces and parterres (level parts consisting of flower beds) were embroidered with statues, urns, terraces and mounds.

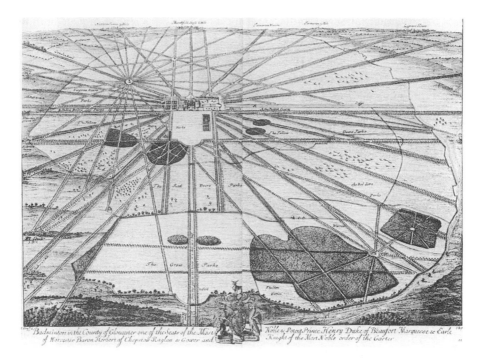

Figure 5.1 *Map of a landscape park in the 'forest style', Badminton, Avon*

Spectating and the ownership of land

As shown previously, mathematics, geometry and practical observation placed the spectator at the centre of their universe, paralleling the revolution in scientific thinking that placed humans as the focus of a universe of dead matter set in mechanical motion. The development of perspective in both Italy and the Low Countries placed the spectator at the centre of worlds that they themselves had created or had commissioned as patrons. The framing of views in the act of picture making, just like the modification of the landscape itself to create gardens, or drain and reclaim land, required a combination of practical and aesthetic judgement. Together these contributed to the sense that humans could and should recreate the world to satisfy their own ends. Thus physical spaces and their representations are transformed into the property of the individual detached observer. To John Berger (1972, 16), perspective was:

> like a beam from a lighthouse – only instead of travelling outwards, appearances travel in. . . . Perspective makes the single eye the centre of the visible world. Everything converges on to the eye as the vanishing point of infinity. The visible

world is arranged for the spectator as the universe was once thought to be arranged for God.

Just like maps, making landscapes on the ground and making images of them in pictures and literature were important parts of the process whereby nature was turned into an object that could be owned, captured, exploited, bought and sold.

In seventeenth-century England, the comparative social stability that followed the restoration of the Stuart monarchy in 1660 saw the expansion of trade and increased wealth for the landed aristocracy. With the great landowners now feeling more secure in their possessions, they directed a growing proportion of their new disposable income to fund the establishment of a new visual order in the landscape, with the construction of new houses, large parks and grand avenues. In addition, most large estates were 'entailed', meaning that property ownership was tied up in such a way that the owner was obliged to pass on the inheritance more or less intact to the next generation. Under such arrangements, it was difficult for estates to fragment and for owners to dispose of land, for example, to cover a gambling debt. Landowners were thus encouraged to consider themselves as *stewards*, having trust of the family inheritance for posterity, and to think long term. Estates grew in size, with outlying lands sold off and holdings consolidated to manage agriculture more efficiently. Williamson (1995, 14) noted: 'it also enabled large landowners to concentrate their political influence, and to increase the opportunities for manipulating the appearance of the landscape to display the extent of their possessions'.

From the 1660s, the Dutch landscape painters became popular with the English landowning aristocracy. Artists such as Johannes Kip and Leonard Knyff adapted the Dutch landscape tradition to the requirements of English aristocratic patrons by painting expansive panoramas of landed estates known as 'Prospects'. As Rosenthal (1982, 22) observed:

> As the Stuart ruling classes liked to perpetuate their own images in portrait, they also liked to see pictorial descriptions of their houses and estates, the Prospects. It is a species of landscape often in the form of a bird's eye view, and one that was usually painted in a scrupulously descriptive style. . . . In these paintings the house, usually through the simple device of being put in the middle of the composition, comes over as the 'centre' of the particular small universe in which it is sited. The long radiating avenues (such as those at Hampton Court) are a pictorial metaphor for the social reality of the influence and power that centred on the great house itself.

Prospect views were both a documentary account, making a detailed visual record of the landed estate, and an opportunity to display the size and grandeur of the estate in an easily understood visual form. They conspicuously presented property as private wealth. Their appearance in England coincided with the development of other important forms of representing land as property on the landed estate, such as the estate survey, the detailed estate map and more sophisticated forms of economic accounting (see Figure 5.1).

Figure 5.2 Bifrons Park, Kent *(c. 1705–10), attributed to Jan Siberechts (1627–1703).*
Courtesy of Yale Center for British Art, Paul Mellon Collection, New Haven, CT

Figure 5.2 shows a view of Bifrons Park in Kent at the start of the eighteenth century attributed to Jan Siberechts (1627–1703). One of the founders of British landscape painting, Siberechts was invited to England in 1670 by an English duke visiting Flanders and remained permanently in England, painting prospect views of country houses. Points of interest in this picture include:

- The presence of the hunting party in the foreground established ownership of the scene unfolding behind them.
- The house with its newly improved garden grounds and adjacent regular enclosed fields contrasted with the wilder 'untamed' wooded landscapes in the foreground and the far distance.
- The trees and irregular ground at either side of the foreground acted as coulisse (see Chapter 4), framing the view and guiding the eye towards the centre of the picture.

- The factual depiction of gardens, barns, houses, fields and woodland showed the estate as a functional agricultural unit and made pictures such as these into a kind of pictorial map.
- Some agricultural workers could be seen labouring but the scale was such that the eye was directed over the top of them so that they appeared 'invisible' even though they were present in the picture. Visual suppression of the evidence of work made the designed landscape and the social order that created it appear to have arisen spontaneously and naturally (Barrell, 1980).
- The village nestled in the shadow of the hill and shaded by trees. The church in the village was a secondary focus. The family hunting party in the foreground forms a visual link between these two parts of the picture, while itself enjoying the protective shade of the trees. The use of trees to shade various elements of the image may have symbolised the benign paternalism associated with the idea of stewardship.

Taken together the elements of the picture suggest harmonious and natural relationships between the authorities of aristocracy and church, landowner and villager, culture and nature.

Landscape design as picture making

In eighteenth-century England, designed elements in the rural landscape, particularly country houses and their surrounding landscape parks, reflected the political split within the aristocracy and gentry between Whigs and Tories. The Whigs included many of the largest landowners, those with important interests in finance and commerce whose values were associated with 'progress' and religious toleration. The Tories, at least in the early decades of the eighteenth century, often supported the Stuart succession and were associated with more traditional, conservative values. These included High Church Anglicanism, minimal involvement in trade and more paternalistic attitudes towards the lower classes (Whyte, 2002, 80). The Whigs had become the dominant political force, largely because they supported the changes in monarchy that resulted in the accession to the throne of Protestant William of Orange in 1688, the constitutional settlement of 1689, and the accession of the Hanoverian King George I in 1714. These political differences were given visual expression in the designs of houses, gardens and landscape parks. The publication of Colin Campbell's book *Vitruvius Britannicus* (1715–25) stimulated interest in neo-Classical architecture in the style of Palladio (see Chapters 4 and 8). The English, as noted in Chapter 4, had come to think of themselves as the inheritors of Western Classical and Renaissance culture and Palladianism became their chosen style. Alongside architecture, the Whig aristocracy and their client artists and writers were strongly influenced by Italian

Renaissance art, particularly the landscape painters Claude Lorrain (known as Claude, 1600–82), Nicholas Poussin (1593–1665), Gaspard Dughet (also called Poussin, 1615–75), Salvator Rosa (1615–73) and the heritage of Classical learning on which they drew. Among these, interpretations of the Classical pastoral in the work of Claude were a dominant influence. Knowledge of European and particularly Italian art and culture was fostered by the 'Grand Tour', the lengthy tour of key sites of European culture primarily undertaken by the sons of the aristocracy in the eighteenth century (Hibbert, 1974; Black, 1992; but also see Dolan, 2002).

Visitors avidly purchased the works of Claude and other seventeenth-century Rome-based artists and brought them back to Britain by the hundred (Whyte, 2002, 89). Back in England, surrounding oneself with the trappings of Italy, the Renaissance and Roman civilisation, became the hallmark of the educated gentlemen. It indicated political allegiance and symbolised an interest in good estate management and the improving agricultural practices of the landowner as *steward*. Whig aristocrats and their artistic circles read Virgil's poetry and identified themselves with the myth of educated rural tranquillity that they found celebrated in the Classical pastoral (see Chapter 4). The reinterpretations of this idyll in the paintings of Claude and others formed a model from which the aristocracy redesigned their estates' landscapes to flatter their self-image.

Figure 5.3 Landscape with Hagar and the Angel *(1646) by Claude Lorrain (c. 1604–82). Courtesy of National Gallery, London*

Figure 5.3 shows a painting by Claude Lorrain entitled *Landscape with Hagar and the Angel* (1646) bought by the landowner, patron, artist and Grand Tourist Sir George Beaumont (1753–1827). It was so important to him and so influenced his own approach to landscape that he is reputed to have carried it with him when travelling away from home. He donated this picture, along with the rest of his collection, to help found the National Gallery in London in 1823, but asked to keep this picture by him until his death. Points to note about it include:

- It employed principles of balanced landscape composition developed earlier in the seventeenth century, often with trees as a frame at the edge. Buildings punctuate the image and act as distance markers. The foreground often includes figures from a story.
- The individual items in the picture stem from observation, but their selection and ordering within the picture are designed to idealise a pastoral conception of nature.
- Golden sunlight, typical of Claudian landscapes, suffused the whole image. In England, merchants sold pieces of yellow glass called Claude glasses so that, by looking through one, observers might recreate this effect when viewing their native landscapes.
- It is painted in the Classical 'picturesque' style, the rough undulating foreground, variegated foliage and buildings that are all rather old and careworn giving the picture a sense of genteel decay. This became a designed feature of many landscape gardens and, later in the century, a highly valued aesthetic characteristic in the British landscape.
- Though the figures adopted dress typical of any representation of the Classical pastoral, this image recounted the story of the servant girl Hagar from Genesis 16. Hagar quarrelled with Abraham's childless wife Sarah and ran away. An angel met her in the wilderness and told her that the child Ishmael would found a great tribe. Meanwhile she should return to Sarah.
- Here the Classical Arcadia fused with the Christian idea of the Garden of Eden. The two pastoral traditions formed a powerful ideological force uniting the two main strands of Western culture.

Claude's paintings suggested perfect harmony between humans and nature. They represented an ideal place, a mental refuge from the real world and reflected the enduring desire in the Western imagination to escape from both society and nature to a comfortable and secure setting. Eighteenth-century England's land-owning gentry effortlessly adopted the idea of refuge, which Claude depicted for his patrons. His trees created pleasant shade for an outing; his peasants gave spectators the illusion of pastoral paradise, not of toiling for a living. Picturesque ruins and well-washed peasants were recognisable as being associated with noble Romans, a Classical association that, as noted earlier, the English were eager to accept

(Crandell, 1993, 98). The effect totally charmed the English gentry: 'There was something in Claude's gentle poetry, in his wistful glances at a vanished civilisation and in a feeling that all nature could be laid out for man's delight, *like a gentleman's park*, which appealed particularly to the English connoisseurs of the eighteenth century' (Clark, 1976, 138–9).

Figure 5.4 shows a map of Rousham Park, near Oxford. It was designed by William Kent (1674–1748) for General James Dormer between 1738 and 1740. Kent has been called the 'Father of English Landscape Gardening' and Rousham is the best and most complete surviving example of his landscaping work (Batey, 1982a, 1982b; Jacques, 1983, 15–42). While it is small-scale, Rousham is perhaps the quintessential English eighteenth-century landscaped garden, with its rolling green lawns, clumps of sturdy oak trees, deer and cattle grazing peacefully on luxuriant meadowland. As 'Le Jardin Anglais', the style was recognised and copied throughout the world (see Mosser and Teyssot, 1991; Woods, 1996).

KEY
 1. Rousham House
 2. The Bowling Green
 3. Lion and Horse (statue by
    P. Scheemakers)
 4. Seats designed by W. Kent
 5. Terrace: Praeneste
 6. Dying Gaul (statue by
    P. Scheemakers)
 7. Arcade designed by W. Kent
 8. Octagonal Pool
 9. Upper cascade with Venus and Cupids
10. Site of upper ponds
11. Serpentine water walk
12. Cold Bath
13. Temple of Echo by W. Kent and
    W. Townsend
14. Gothic seat by W. Kent
15. Palladian gate
16. Gothic bridge over Cherwell
17. Statue of Apollo
18. Long Walk
19. Lower cascade
20. Ampitheatre by J. Bridgeman
21. Pyramid by W. Kent
22. Gothic seat
23. Walled garden
24. Dovecot
25. Church
Arrows indicate the preferred direction of
the garden circuit

Figure 5.4 *Map of Rousham Park, near Oxford, England*

Source: Based on a map in N. Pevsner and J. Sherwood *Oxfordshire* ('The Buildings of England Series') (Penguin, 1974)

Kent trained as a sign painter, but soon started to paint pictures. In 1719, the rich and powerful third Earl of Burlington (1694–1753) brought him to London where he designed furniture, gardens and buildings for Burlington and his associates. Like many of his contemporaries, Burlington had taken the Grand Tour of Europe and visited Italy, where he studied the art and architecture of Ancient Rome and the Italian Renaissance. He particularly liked the paintings of Nicholas Poussin and Claude Lorrain and the architecture of Andrea Palladio and the Roman architect Vitruvius. Along with a number of 'enlightened' and 'educated' friends and acquaintances, such as the writer Joseph Addison and the poet Alexander Pope, he helped to establish the enthusiasm among the eighteenth-century aristocracy for all things Italian.

Kent's introduction to General Dormer came through the latter's friendship with Alexander Pope. Charles Bridgeman, who was Burlington's first landscape gardener, had undertaken the previous landscaping at Rousham in 'forest style', that is, with formal geometric rides and avenues cut through trees (Hadfield, 1977, 29–36; see Figure 5.1). William Kent designed gardens to look like the pictures of Poussin, Claude and Salvator Rosa. His landscapes combined temples, grottos, amphitheatres and statues as one might find in ancient Rome with the woodland glades, streams, pastures, craggy hills and well-managed farms that one could read about in the poetry of Virgil and Horace. Using the compositional techniques derived from painting, Kent's picturesque Claudian landscape, framed by coulisses, led the eye via a series of highlights to a vanishing point at the horizon. This became a formula for painting and landscape design, repeatedly copied in eighteenth-century England.

Figure 5.5 provides a general view of the scene. Beginning at the house, the primary view was determined by a wide bowling green stretching away and terminating where the land slopes steeply to the River Cherwell. This point was marked by a centralised sculpture of a horse being attacked by lions. Woodland coulisses framed this perspective, their role emphasised by ornamental seats at the corners of the lawn. Beyond them, the slope has been hollowed out to exaggerate the drop to the river, cutting out the middle ground and extending the view into the distance. In the middle distance, Kent designed a mill house with stepped gables and flying buttresses to resemble a ruined chapel and, on the horizon, a sham arch of three tall arched lights and a decorated gable to suggest a medieval monastic ruin.

However, as Cosgrove (1984, 201) showed, this is a picture that can only be read fully by those who understand its symbolism. The Gothic eyecatcher stands significantly outside the garden in the unadorned English countryside. Gothicism was a widespread and attractive literary genre that argued that a revival of Western civilisation had emerged from the union of the opposing principles of Roman

Figure 5.5 *Rousham Park, view from the house*

authority and Gothic liberty. Its ideals matched those of British nationalism and the constitutional form that had emerged from the events of 1689 (see above). Kent's use, therefore, of both Classical Roman and Gothic allusions in the Rousham landscape can be read as a statement of perfection in the English land-scape – a reflection of the legitimacy of those, like General Dormer, who ruled over it. In the garden itself the owner revealed his own taste and educated sensibility: the eye was led out over a scene associated with the Roman *campagnia*, but into an unmistakably English landscape, consciously alluding in its architectural ruins to the traditional freedoms of Britain (Cosgrove, 1984, 203). By using a complex symbolic language of allegory and allusion, English landscape designers used nature to turn Italian design into English patriotism.

Enclosure and the counter pastoral

Much of the eighteenth-century British countryside testified to the domination of the powerful over the powerless (Whyte, 2002). By the turn of the nineteenth century, not just the countryside and their parks but the whole fabric of the countryside embodied the values of the landed elite, their aesthetic preferences, their economic interests and their leisure pursuits. The eighteenth-century Enlightenment, with its emphasis on order and the control of nature, became stamped on the British landscape. Though not affecting all the country, the principal expression of the new regulated productive landscape was Parliamentary enclosure. The period between 1770 and 1815 witnessed the most intensive activity, affecting more than seven million acres (2.9 million hectares) of land.

Parliamentary enclosure transformed open-field arable land in a wide belt of country running from south-central England to east Yorkshire, as well as extensive areas of open heath and common and large blocks of upland waste (see Figure 5.6). In lowland areas, these landscapes were characterised by regular patterns of square or rectangular fields surrounded by hawthorn hedges and accessed by wide, straight roads, dotted with new farmsteads (Williamson, 2002, 125–9; Reed, 1990, 308–10). In upland areas the grids of stonewalls, constructed in a similar way in each area, created 'ever more regimented landscapes' (Whyte, 2002, 75). In the process the customary rights of villagers, substantially unchallenged for generations, were eroded. Animals could no longer be grazed on the commons, game and firewood could no longer be collected from woodlands, and the poor and the old could no longer pick up 'gleanings' from the fields in order to supplement their diets. By the mid-nineteenth century, so-called clearances in the Highlands of Scotland resulted in substantial emigration. The notorious Black Acts (1727) set the tone for the century by making it a hanging offence even if only suspected of poaching – specifically, appearing with a blackened face within the vicinity of a game reserve.

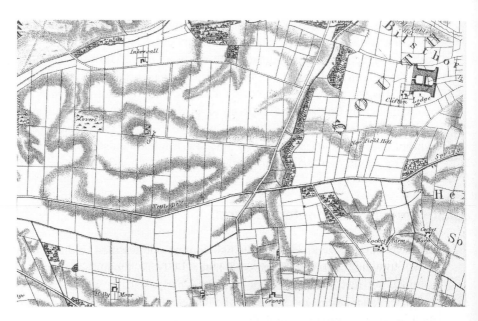

Figure 5.6 *Map of regular fields in an eighteenth-century enclosure landscape, north Nottinghamshire*

The new inequalities became highly visible after 1750. A period of economic expansion, rising agricultural prices and rising rents greatly benefited the landowners. This resulted in poverty and unemployment as population growth began to outstrip the number of jobs available in villages, the value of real wages fell and inflation set in. The situation in rural communities deteriorated still further in the last decades of the century when the blockade of Britain during the Napoleonic Wars caused food prices to rise dramatically. Williamson (1995, 101) noted:

> Throughout the second half of the century landlords were keen to cash in on the opportunities offered by demographic recovery and economic expansion, and seem in general to have managed their estates with a greater eye to efficiency and profits than before. They were eager, above all, to amalgamate small family farms into larger, more productive units. In a multitude of ways the old rural order was changing.

During this period under the influence of the landscape designer Lancelot 'Capability' Brown, park design became increasingly large scale and austerely abstract, reducing the 'picturesque style' of William Kent to an extensive canvas of rolling green lawns, sinuous lakes and clumps and belts of woodland trees (see Figure 5.7). Though huge expense was not the norm (Williamson, 1995, 108), under Brown's influence park landscaping could be very expensive. Amongst Brown's more costly creations were Navestock (Essex) £4,500, Thorndon (Essex) £5,000, and Blenheim (Oxfordshire) and Langley (Buckinghamshire) together £21,000. Increasingly such exercises required the closure and re-routing of public roads and the removal, and sometimes repositioning, of entire villages from their medieval locations close to manor and church. Notable examples of this included Nuneham Courtenay (Oxfordshire) and Chatsworth (Derbyshire). Williamson (1995, 100) characterised this as a 'landscape of exclusion'.

During this period the elite aesthetic practice of landscaping was challenged by a counter-pastoral representing the views and interests of ordinary agricultural labourers in poetry, the same medium as in the original Classical pastoral of Virgil and Horace (Williams, 1973). Important works in this genre included *The Thresher's Labour* (1730) by Stephen Duck (1705–56) who began life as a farm labourer in Wiltshire, *The Deserted Village* (1770) by Oliver Goldsmith (1728–74) and *The Village* (1783) by George Crabbe (1754–1822), a Suffolk parson and doctor. Though not a labourer himself Crabbe championed the cause of the agricultural labourer against the landowners. In this poem, agricultural labourers were slaves being worked to death – a quite different image from the pastoral ideal of happy shepherds and poetic ploughboys.

> Or will you deem them amply paid in health,
> Labour's fair child, that languishes with wealth?
> Go then! and see them rising with the sun,
> Through a long course of daily toil to run;
> See them beneath the dog-star's raging heat,
> When the knees tremble and the temples beat;

> Behold them, leaning on their scythes, look o'er
> The labour past, and toils to come explore;
> See them alternate suns and showers engage,
> And hoard up aches and anguish for their age;
> Through fens and marshy moors their steps pursue,
> When their warm pores imbibe the evening dew;
> Then own that labour may as fatal be
> To these thy slaves, as thine excess to thee.

Figure 5.7 *Drawing of a Capability Brown style landscape by Thomas Hearne (1744–1817)*

William Cobbett, whose *Rural Rides* (1821) celebrated agricultural improvement, recognised the value of commons and wastes as a resource. Cobbett was sometimes critical of enclosure and argued that the expenditure did not make up for the losses. Rabbit warrens, for example, were often cleared away by enclosure even though they formed a significant economic resource for villagers. In a *Memoir* written in the 1820s remembering the loss of a Northumberland common of the 1780s, Thomas Berwick (1755–1828), noted:

> On this common – the poor man's heritage for ages past, where he kept a few
> sheep, or a Kyloe cow, perhaps a flock of geese and mostly a stock of bee-hives

– it was with infinite pleasure that I long beheld the beautiful wild scenery that was there exhibited, and it is with the opposite feeling that I now find all swept away. Here and there on this common were to be seen the cottage, or rather the hovel, of some labouring man, built at his own expense, and mostly with his own hands; and to this he always added a garth and a garden, upon which great pains and labour were bestowed to make productive. . . . These various concerns excited the attention and industry of the hardy occupants, which enabled them to prosper, and made them despise being ever numbered with the parish poor. These men . . . might truly be called – 'a bold peasantry, their country's pride'.

(quoted in Williams, 1973, 100)

Exercise 5.1

Read the above quote, then look back to the example of the Rousham Park landscape discussed earlier in this section.

Compare the way each represents nature as a source of pride and identity.

List three similarities and three differences.

Your list may include the following similarities. At both Rousham and on Berwick's Northumberland common:

- Landscape is a source of pride and identity.
- Human agricultural activity and garden design is an expression of pride and identity.
- History and memory are built into our involvement with land making landscape a source of socially meaningful symbolic value.
- The control of nature and its improvement is a mark of human status.

Differences might include:

- At Rousham pride in landscape involves standing back and observing the view.
- In Berwick's account pride in landscape is found by being actively engaged with the land on a practical basis.
- The Rousham landscape makes direct reference to the values of the elite cultures of Italy and the elite politics of the dominating Whig aristocracy.
- Berwick's Northumberland common makes direct reference to the values of an independent and self-reliant agricultural labourer.
- Rousham represents an elite conception of landscape.

- Berwick's Northumberland common represents a labourer's conception of landscape.

Yet, as with Crabbe, the author did not represent the authentic voice of the dispossessed, even though his *Memoir* provided insight into real suffering. Berwick was an artist and engraver best known for his illustrations of birds and animals (Osborne, 1970, 132–3). At best, Berwick and Crabbe showed the new found empathy with the 'lower orders' and with nature that became increasingly characteristic of middle-class and town based populations in the nineteenth century.

Romanticism and reaction, 1770–1830

The typical ideas from the Romantic Movement highlighted in this part of this chapter include:

- reaction against industrial revolution;
- appreciation of nature for its own sake;
- the first moves to protect nature from the effects of human activity;
- the revival of religious and mystical perspectives on nature derived from Christian, classical and pagan thought;
- the idea that direct experience of nature brings physical and mental benefits;
- the belief that individual subjective experience of nature is as important (if not more important) than science.

Romanticism exerted a profound influence in nineteenth-century Europe and North America, with its legacy continuing through into the twentieth and present centuries. A reaction to Enlightenment thinking, Romanticism fed on the fears and misgivings accompanying large-scale industrialisation and urbanisation (see also Chapter 7). This period saw a rapid growth in scientific exploration and technological achievement that, in turn, was associated with the advent of the Industrial Revolution and accompanying disruption of traditional social patterns and values. Romanticism, however, was not something completely new. It drew on both Classical and Christian ideas of nature, while grounding its visual styles in Italian and Dutch Renaissance landscape art. Romanticism was an artistic and intellectual movement, with expressions in literature, music, painting and drama. At the same time, it was not just a set of ideas unrelated to what was happening in the material world, since Romanticism was a clear reaction *against* the material changes occurring in society (see also Chapter 7).

Romanticism has numerous strands and its ideas have been reworked on many occasions (Pepper, 1996, 188–92), but some of its main characteristics were as follows:

- As a moral code for living, the Romantic Movement championed aesthetic ideas of truth and beauty over practical concerns with usefulness and material value.
- Romantics celebrated the freedom of the individual as expressed in feelings and emotions as uniquely experienced by each person. This contrasted with the search for universal laws and generalisations about nature and people, which enabled scientists to try and predict behaviour.
- Many Romantics, notably Thoreau and Ruskin, were interested in the scientific field study of nature, yet refused to accept the way that science abstracted aspects of nature from the whole. Rather, Romanticism looked to holistic and organic conceptions of nature characteristic of the Medieval and Renaissance periods.
- Romanticism effectively criticised commonsense and mainstream social and political values and was compatible with radical and revolutionary political views. The early years of the Romantic Movement coincided with the American and French Revolutions, while later nineteenth-century Romanticism was associated with the anti-industrial socialism of William Morris and others (see Thompson, 1997).
- Subjective knowledge of, and oneness with nature – particularly as expressed through art – was a superior form of knowledge to that of objective, coldly calculating science. Romantics cherished fantasy, imagination and unrepressed depth of feeling. Spontaneity, inner truth and the unique point of view of the artist were fundamental sources of value.
- Romantics largely opposed the form of 'progress' represented by industrialising society. Instead they found value in the simplicity and honesty of nature rather than the sophistication and complexity of civilisation.
- Following this, they celebrated more traditional forms of society both in the past and in less developed regions where they imagined a simpler life to be both more honest and closer to nature in contrast to the corruption of modern urban society.

Tourism, the sublime and the picturesque

One consequence of Romantic veneration of 'wild' landscapes has been to leave the legacy of an aesthetic of landscape that values mountain, moorland and marsh for their pictorial qualities in ways unknown to previous generations. Until the later eighteenth century, nature was popularly imagined as inhospitable; associated with impenetrable forests, fearsome wild animals, unscalable mountains, gaping ravines, hostile demons and unhealthy peasants (Tuan, 1993, 60–1). Even areas among the first to be reimagined as 'beautiful' scenery had been regarded as turbulent and unmanageable only a few decades earlier. In the early years of the eighteenth century, Daniel Defoe viewed the English Lake District as 'the wildest, most barren and frightful' of any that he had visited (Nicholson, 1978, 25). Dr Samuel Johnson dismissed the central Scottish Highlands as having little to detain the traveller: 'An eye accustomed to flowery pastures and waving harvests is astonished and repelled by this wide extent of hopeless sterility' (Chapman, 1924, 34). However, during the course of the eighteenth century the rage for mountain scenery took firm possession of 'the holiday seeking public'. John Byng, the diarist Viscount Torrington, climbed Cader Idris in Wales in 1784 with a guide who had been taking tourists up the mountain for the previous 40 years. By 1800 the poet Coleridge, who had helped to make the Lake District popular, complained that 'the Lakes were alive and swarming with tourists for a third of the year' (quoted in Thomas, 1971, 260). By the late eighteenth century the appreciation of nature, and particularly wild nature, had been converted into a sort of religious act. Nature was not only beautiful; it was morally healing.

Wilderness came to occupy a special place in people's affections. The *Oxford English Dictionary* gives the origins of the term in Old English (*wild(d)éornes*) and Middle Dutch (*wildernisse*), as literally meaning the 'place of wild animals' (Nash, 1982, 4) and it still resonates with Biblically inspired connotations of places filled with evil and temptation. Yet the God-forsaken wilderness was about to become sacred space; places of scenic grandeur characterised by solitude and tranquillity, in which the individual could gain personal enrichment through closeness to nature and to God. As Thomas (1971, 260) suggested:

> The value of wilderness was not just negative; it did not merely provide a place of privacy, an opportunity for self-examination and private reverie (which was an ancient idea); it had a more positive role, exercising a beneficent spiritual power over man [*sic*]. . . . The feeling of awe, terror and exultation, once reserved for God, was gradually transposed to the expanded cosmos revealed by the astronomers and to the loftiest objects discovered by explorers on earth: mountains, oceans, deserts and tropical forests. The inhabitants of mountain areas ceased to be universally despised for their barbarism; instead they were praised for their innocence and simplicity.

By the beginning of the nineteenth century, the Lake District along with the Welsh mountains, the Scottish Highlands and England's Peak District were popular destinations for middle-class tourists carrying newly written guide books and looking for spectacular scenery, views they could draw and paint (Figure 5.8) and close encounters with raw nature (Urry, 1995; Darby, 2000, 83–91). Macnaghten and Urry (1998, 114) called this the 'spectacularisation of nature', arguing that it coincided with the development of various visual theories and codes, especially the sublime and the picturesque, that enabled the more terrifying aspects of nature to be reinterpreted as part of a meaningful aesthetic experience.

Figure 5.8 *Amateur artists on a sketching trip in Wales by Paul Sandby (1730–1809).*
Courtesy of Yale Center for British Art, Paul Mellon Collection, New Haven, CT

The sublime

The aesthetic of the sublime derived from the writings of Immanuel Kant (1724–1804) on aesthetic judgement. In the British context the most important single work is *Philosophical Enquiry into the Origin of Our Ideas of the Sublime and the Beautiful* by Edmund Burke (1729–97). Burke urged his readers to see the sublime in that which was vast, rugged, dark, gloomy or excited ideas of pain and danger. From such scenery 'came ideas elevating, awful and of a magnificent kind' (Womersley, 1998). The rugged rocks, precipices and gloomy mountain torrents had become the yardstick of aesthetic experience. For some Romantics their dark soaring masses expressed the immensity and infinity associated with God. To experience the sublime was to experience simultaneously terror and delight (Macnaghten and Urry, 1998, 114). The artistic idea of the *sublime*, then, appreciated the 'harmful' as well as the safe (Pepper, 1996, 194). The combination of witnessing danger from a position of comparative safety, which characterised the sublime, is perhaps evident today in the popularity of theme park rides, horror films and extreme sports.

The beginnings of spectatorship at home from the 1760s onwards was connected initially with an expanding economy, a degree of domestic stability, improving roads, more money and leisure for an expanding class of landowners and smaller gentry and an increasingly important urban middle class. Such people could not aspire to the Grand Tour. Poets like Thomas Gray (1716–71) had already travelled through the Alps and brought the new ways of seeing to bear on the wild and remote areas of the British Isles. Gray had crossed the Alps in 1739; he visited the Lake District in 1769 and published his journal account of the visit in 1775 – one of the first accounts for tourists written adopting the language of picturesque landscape (Andrews, 1999, 11; Whyte, 2002, 98). In addition, the rise of a confident and wealthy urban-based, middle class resulted in a new market for works of art not tied to the aristocratic social codes of landed property and agricultural improvement.

However, the new ways of seeing were firmly grounded in the past. The *sublime* and the *picturesque* were tied to the old pictorial conventions of the seventeenth-century Italian artists who had inspired eighteenth-century landscape design (Andrews, 1999). The Italian painter Salvator Rosa in particular acquired a reputation for wild, turbulent landscapes. Grand Tourists bought originals, painted copies, and prints of his work, which became increasingly familiar in the collections of the great houses in northern Europe in the eighteenth century. Among the cultivated classes, Rosa's name became almost proverbial for the terror induced by awesome mountain scenery. When the aristocrat Horace Walpole crossed the Alps with Thomas Gray in 1739, he wrote a letter to a friend exclaiming 'Precipices, mountains, torrents, wolves, rumblings, Salvator Rosa'. Walpole championed a picturesque gothic style forming an important link between Classicism and Romanticism in the eighteenth century. Meanwhile Johann Ludwig Aberli, a printmaker in Berne in the late eighteenth century who specialised in views of Swiss scenery, recalled a journey into the Alps made in 1774: 'On our travels it sometimes happened that both of us would cry out at the same time *Salvator Rosa! Poussin! Saveri! Ruisdael! Or Claude (Lorrrain)* according to whether the subjects before our eyes reminded us of the manner and choice of one or other of the masters named' (Andrews, 1999, 130–1; see also Hauptman, 1991, 13–15).

In addition to conscious picture making closely associated with the picturesque view of landscape, tourists and travellers also learned to appreciate nature as a flow of experience rather than a carefully composed scene, as indicated by Walpole when describing his journey with Gray to their mutual friend Richard West. This way of perceiving nature is more closely associated with the aesthetic of the sublime. No longer looking for control and order in nature, tourists sought out experiences for their intensity and overwhelming power:

But the road, West, the road! Winding round a prodigious mountain, and surrounded with others, all shagged with hanging woods, obscured with pines, or lost in clouds! Below, a torrent breaking through cliffs, and tumbling through fragments of rocks! Sheets of cascades forcing their silver speed down channelled precipices, and hasting into the roughened river at the bottom! Now and then an old foot-bridge, with a broken rail, a leaning cross, a cottage, or the ruin of an hermitage! This sounds too bombast and too Romantic to one that has not seen it, too cold for one that has. . . . We stayed there [Grande Chartreuse] two hours, rode back through this charming picture, wished for a painter, wished to be poets!

(cited in Andrews, 1999, 130)

Exercise 5.2

Figure 5.9 shows a painting called *An Avalanche in the Alps*. It dates from 1803 and is by Philippe Jacques de Loutherbourg (1740–1812). Compare this picture with the country house prospect of Bifrons Park (Figure 5.2).

Make a list contrasting the relationships between humans and nature in these two images.

Figure 5.9 An Avalanche in the Alps *(1803) by Philippe Jacques de Loutherbourg (1740–1812). © Tate, London 2003*

Your list should include some of the features cited in Table 5.1. Though born in Strasbourg, de Loutherbourg worked in England from 1771, where he earned his living as a stage designer while also becoming known for his large set piece dramatic pictures. These included his paintings of battle scenes and his images of the iron works at Coalbrookdale by night lit by the glow of the furnaces. He also invented a mechanical moving panorama show called the Eidophusikon, which showed storms, running water, fire and sundry natural disasters through the manipulation of light and movable images (Ceram, 1965). With regard to this picture, Hauptman (1991, 98) suggested that: 'However savage or sublime to the English eye, no artist at the time was concerned visually with the demonic danger of the Alps or the persistent menace of natural catastrophe as was de Loutherbourg through the exhibition of this remarkable canvas.'

Table 5.1 *Comparison between the paintings of 'Bifrons Park' and 'Avalanche in the Alps'*

Bifrons Park	Avalanche in the Alps
High viewpoint	Low viewpoint
Nature benign and unthreatening	Nature is an all powerful and uncontrollable form
Evidence of human control of nature (e.g. fields, farms, managed woodland)	Evidence of human frailty in the face of nature (e.g. wrecked bridge)
Humans prominent in the foreground	Human figures very small and visually off centre

The picture exemplifies what Hauptman called disaster iconography, which found particular favour with safe and highly urbanised elite classes of late eighteenth-century Paris and London. Its popularity can also be witnessed in the success of the Eidophusikon in London – one of a spate of optical novelty theatres – where de Loutherbourg made dramatic entertainment out of representing natural processes. Though there are literary records of alpine avalanches dating back to the middle of the sixth century, it was not until the nineteenth century that descriptions of them became commonplace in travel literature. Only with the new aesthetic of the sublime were painters able to find a way of representing overwhelming natural forces without using Classical allegory or overt religious symbolism. Further things to note in this picture, include:

- The way in which the elements of the picture were dramatised to tell a story.
- The small size and displacement of the human figures from the centre of the image.

- The overwhelming energy and scale of physical nature, its dynamism and movement (compared, for example, to the controlled and ordered nature of the landscape prospect). This is symbolised by the strong diagonal structure of the composition, with diagonal movement frequently used to symbolise power and aggression.
- The theatrical poses of the human figures; one man appears to call alarm, another seems to be making his escape, a third prays, as if showing the variety of individual and personal ways in which humans react to nature as an overwhelming force.
- The fragility of human life and the inadequacy of human strength and ingenuity against the overwhelming forces of nature: symbolised by the bridge, which is rather insubstantial and easily swept away by the raging torrent.
- The figures were both safe and threatened as they sheltered under the overhanging rock. This paralleled the viewers' experience of this image, able simultaneously to imagine themselves part of this drama while knowing that they were safe as mere spectators.

The position of being both spectator and potential victim-participant is crucial to the full experience of the sublime (Andrews, 1999, 135). The feelings are both destabilising and reassuring, together making the sublime pleasurable and exciting through its dramatic tension. After the philosopher Jean-Jacques Rousseau (1712–78) made a trip into the French Alps, he wrote: 'I must have torrents, fir trees, black woods, mountains to climb or descend, and rugged roads with precipices on either side to alarm me'. Nevertheless, he admitted later in the same passage to not relishing the challenge of real danger. To be enjoyable his sense of alarm required a measure of security: 'The road has been hedged by a parapet to prevent accidents, and I was thus enabled to contemplate the amusement in these steep rocks as they cause a giddiness and swimming in my head which I am particularly fond of, provided I am in safety' (Coates, 1998, 133–4).

Just as the sublime drew consciously on the work of Salvator Rosa, so the 'picturesque' found its inspiration in the paintings of Claude. Though William Kent's style of garden design from the 1740s is called 'picturesque', *picturesque tourism* developed from the 1770s onwards, giving rise to the formalisation of its aesthetic rules in theoretical treatises and guidebooks. Its greatest publicist was the Reverend William Gilpin (1724–1804). Originally from Cumbria, he took holy orders and pursued a successful career as a headmaster in the south of England before early retirement to a parish in Hampshire gave him sufficient leisure to undertake a series of tours and write them up. Gilpin's accounts of picturesque tours to the Wye Valley (1770), the Lake District (1772), the Scottish Highlands (1776) and other parts of Britain circulated privately in manuscript form for some years before they were published (Ousby, 1990; Whyte, 2002, 98–9).

The picturesque

To Gilpin, a 'picturesque' landscape was one that would look good in a picture painted by one of the great neo-Classical artists of seventeenth-century Italy, notably Claude, Poussin, Dughet or Rosa. Gilpin saw the picturesque as a subset of Burke's category of beautiful and sublime. Its qualities included roughness, variety, irregularity and intricacy. Something that was beautiful pleased the eye in its natural state but, with something that was picturesque, its quality could best be appreciated when it was depicted in a painting. Some picturesque theorists also stressed the importance of historical associations in the landscape. Uvedale Price, for example, thought that picturesque scenes needed a time dimension. Pictures including ruins and showing the process of decay linked the picturesque centrally to Romantic thought.

From the 1780s, Gilpin encouraged picturesque tourism and the associated appreciation of landscape by middle-class travellers. There was a ready market for such guidebooks. The new generation of middle-class people were relatively untutored in the visual arts and had the time to spare to view the scenery of Wales, the Lake District or Peak District, but wanted to know what to look at and how to see in much the same way that present-day travellers to exotic locations use Lonely Planet and Rough Guides guidebooks. Picturesque tourism in Britain produced an explosion of descriptive travel literature in the last two decades of the eighteenth century, following the publication in 1782 of Gilpin's tour of the Wye Valley. Although it was very different in character from the sublime, the picturesque's visual codes relied very heavily on the concept of the sublime and its apparent opposite, the beautiful. As the sublime stressed disorder and intense experience, so the picturesque stressed calculated and measured composition. It set spectators at a distance from the landscape enabling them to turn the landscape in front of them into pictures governed by an explicit set of rules. These were published in the guidebooks written by theorists such as Gilpin, who noted in the introduction to his *Observations On The River Wye* (1770):

> The following little work proposes a new object of pursuit; that of not barely examining the face of a country; but of examining it by the rules of picturesque beauty: that of not merely describing: but of adapting the description of natural scenery to the principles of artificial landscape; and of opening the sources of those pleasures, which are derived from the comparison.

Some theorists thought that nature itself held picturesque qualities. Others, including Gilpin, considered that the observer makes nature into a picture by the rigorous application of picturesque principles. Either way, all agreed that the picturesque could be found in a combination of the sublime and the beautiful – a mix of roughness and smoothness, awe, grandeur and submissive tranquillity (see Table 5.2).

Table 5.2 *The sublime, the beautiful and the picturesque*

Sublime	Beautiful	Picturesque
Mountains	Lakes	Combination of roughness and smoothness
Waterfalls	Lawns	
Cliffs and precipices	Low undulating landscape	Variation and variegation
Severe weather	Winding rivers	Shrubs and flowering plants
The power of nature	Benign nature	Highly textured plants and landscapes
Verticality	Horizontal landscapes	
Grandeur	Sparkling brooks	Mixture of old and new
Fear	Brightness	Historical associations
Darkness	Happiness	Controlled composition
Ruggedness	Smoothness	
Savageness	Submissiveness	
'Masculine'	'Feminine'	

One consequence of Gilpin's conception of the picturesque as a formal and relatively objective set of rules applied to nature was that, if the rules were applied in an appropriate manner, almost any scene could be valued as attractive and of artistic value. Both industrial workers and labouring people could now be admitted into the view as long as they contributed to its picturesque qualities. Figure 5.10, for example, shows a painting entitled *Iron Forge at Tintern* (1790) by Thomas Hearne (1744–1817). Hearne was a well-known topographical artist, often engraving his views to allow them to be produced in bulk to serve the growing market for picturesque views. Tintern, in the Wye Valley on the border of England and Wales, was a favourite spot for artists and picturesque tourists because of the ruined Abbey church set within a deeply wooded valley. It was also an active industrial site, with the nearby Forest of Dean an important coal mining region since Roman times. During the Middle Ages, Dean was Britain's premier iron-producing district; in the seventeenth century, it had the largest concentration of charcoal blast furnaces in Britain (Hart, 1971; Fisher, 1981; Coones and Patten, 1986; Ousby, 1990).

Figure 5.10 Iron Forge at Tintern *(1790) by Thomas Hearne (1744–1817)*

Exercise 5.3

Using Table 5.2 as a guide, outline the ways in which the picture shown in Figure 5.10 conforms to picturesque principles.

Your answer should include the following:

- The combination of roughness, smoothness and variegation in texture characterise this as a picturesque view.
- The historical associations of iron working in this area also confirm its picturesque credentials.

Other points to note about this image include the following:

- The overhanging trees, thick undergrowth around the first water wheel, the rough texture of the walls and roof, the foreground scattered with timber, stone and other debris suggest that the site had been abandoned and left to decay. It was likely that the iron forge was in operation during this period and therefore this image demonstrated picturesque decay for visual effect.

- Here working industry was reduced to a visually attractive ruin – viewing industry as if it was already worn out makes it seem less threatening. It enabled the viewer to imagine the restoration of a 'natural' garden landscape after the upheavals of industrialisation. Not surprisingly the picturesque has recently been associated with the heritage industry and the fascination with industrial archaeology that has developed in Britain since the 1960s (see Hewison, 1987).

Nature, introspection and transcendence

An important theme in the Romantic approach to nature was the extent to which contact with 'wild nature' inspired Romantics to try to think beyond everyday material life and contemplate a spiritual realm of existence. This tendency within Romanticism, called *transcendentalism*, came to prominence through the views of American writers and thinkers on nature such as Ralph Waldo Emerson (1803–82) and Henry Thoreau (1817–62). For example, Thoreau said: 'For the religious all wild nature is a manifestation of God (pantheism)' and 'we need the tonic of wilderness . . . we can never have enough of nature' (Pepper, 1996, 200). Transcendentalist Romantics believed that contact with 'wild nature' purified and refreshed people spiritually. This, in turn, became a fundamental tenet of the movement for landscape preservation and the establishment of national parks, in both the USA and the United Kingdom (Worster, 1985, 57–111). As John Muir wrote in 1898: 'Thousands of tired, nerve-shaken, over-civilised people are beginning to find out that going to the mountains is going home; that wilderness is a necessity and that mountain parks and reservations are useful not only as fountains of timber and irrigating rivers but as fountains of life' (quoted in Pepper, 1996, 200).

This way of thinking about nature was novel in its implications for the apprecia-tion of nature and landscape by a mass population rather than to elite groups of artists and writers, yet it reached back to the core of Romantic thinking. The poet Samuel Taylor Coleridge, for example, referred to torrents and cataracts in the Savoy Alps as 'glorious as the Gates of Heaven' in his poem *Hymn before Sunrise, in the Vale of Chamouni* (1802). For his friend William Wordsworth, mountains were 'temples of Nature' because they inspired spirituality and sacredness without specifically referring to the Christian God (Coates, 1998, 135). Schama (1995, 478) described how in addition to their representation as a theatrical stage for the sublime power of nature as seen in Figure 5.9, the Swiss Alps became valued as intrinsically good places where the lifestyle of the inhabitants was seen to suggest a model society.

The representation of mountains as a pure and moral natural environment for human society linked in with both political radicalism and the Romantic celebration of ordinary working lives lived in close contact with agricultural routine. People living in hills and mountains, extracting a hard-won living in difficult circumstances, were represented as sturdy, independent folk; the upholders of liberty and morality. Rousseau, famous amongst other things for his idealisation of the 'noble savage' – the intrinsically good human being living in a state of nature – portrayed the Alps as a barrier separating the honest Swiss from the vices of other nations. The 'honest hunger' of their hills offered the inhabitants only crude iron ore as a resource. Yet this apparently was a source of envy 'for all hardships vanish where liberty reigns and the very rocks are carpeted with flowers' (quoted in Schama, 1995, 481; see Hauptman, 1991, 16–17).

Fascinated by the way in which a deep love of nature grew from an intimate knowledge of place, Wordsworth and other Romantics took to walking excursions and tours as a source of spiritual and aesthetic fulfilment. Anne Wallace (1993; also Jarvis, 1997), for example, showed how walking was *aesthetically* represented in the work of William Wordsworth, through the literary device of the 'peripatetic'. This brought together agrarian values, stressing work and good husbandry, with a Romantic sensibility that cherished proximity to nature. In walking, Romantics found a means of engaging with landscape and nature that involved real physical effort and encouraged local environmental and topographical knowledge. For middle-class poets, writers and painters, the effort of walking stood in place of the labour of working the land, and environmental and topographical knowledge took the place of the farmer's intimate connection with plants, animals, the seasons, and the soil. Tracing the 'peripatetic' through the nineteenth century and finding clear traces in current ideas of psychic and physical health in the Western world, Wallace (1993, 13) maintained that:

> Essays by William Hazlitt, Henry David Thoreau, John Burroughs, Robert Louis Stevenson, and Leslie Stephen, although differing in detail, all argue that the natural, primitive quality of the physical act of walking restores the natural proportions of our perceptions, reconnecting us with both the physical world and the moral order inherent in it, and enabling us to recollect both our personal past and our national and/or racial past – that is, human life before mechanization. As a result, the walker may expect an enhanced sense of self, clearer thinking, more acute moral apprehension, and higher powers of expression.

The poem 'Lines composed a Few Miles above Tintern Abbey, On Revisiting the Banks of the Wye during a Tour (July 13, 1798)' by William Wordsworth is frequently quoted because it revealed a Romantic approach to nature. The poem was composed on a four-day walk to Bristol undertaken with his sister Dorothy and finally written down when they arrived. As already shown, the Wye Valley was an important location for picturesque tourists and the poem recounted Wordsworth's return to a particular spot in the Wye Valley after five years' absence. The following passage described how Wordsworth's conception of nature had changed during this time:

> when like a roe
> I bounded o'er the mountains, by the sides
> Of the deep rivers, and the lonely streams,
> Wherever nature led: more like a man
> Flying from something that he dreads, than one
> Who sought the thing he loved. For nature then
> (The coarser pleasures of my boyish days
> And their glad animal movements all gone by)
> . . .
> The sounding cataract
> Haunted me like a passion: the tall rock,
> The mountain, and the deep and gloomy wood,
> Their colours and their forms, were then to me
> An appetite; a feeling and a love,
> That had no need of a remoter charm,
> By thought supplied, not any interest
> Unborrowed from the eye. – That time is past,
> And all its aching joys are now no more,
> And all its dizzy raptures. Not for this
> Faint I, nor mourn nor murmur; other gifts
> Have followed; for such loss, I would believe,
> Abundant recompense. For I have learned
> To look on nature, not as in the hour
> Of thoughtless youth; but hearing oftentimes
> The still sad music of humanity.

Exercise 5.4

Briefly itemise the evidence from these lines of poetry which suggests Wordsworth's typically Romantic approach to nature.

How has Wordsworth's approach to nature changed during his period of absence?

The beginning of this extract contained an approach to nature that was thoroughly Romantic. Wordsworth described himself as being like a young deer totally engaged and engrossed in nature like an animal spirit lacking reflection and self-consciousness. However, his perspective changed from the idealism of youth to the reflective maturity of adulthood. Between November 1791 and December 1792, Wordsworth was in France supporting the Revolution. Although he remained concerned about social inequalities, he became disenchanted with the French Revolution during the period of 'the Terror' and the suffering and distress that it caused. As a result, he changed his ideas about enforced social reform and with them his perspective on nature. By the end of his life, Wordsworth had become quite conservative and reactionary. This poem was one that marked this transition.

Though Wordsworth has been championed as a proto-environmentalist (e.g. Bate, 1991), it is important to remember the extent to which views can and do change during the course of a lifetime. The Wordsworth that Bate celebrated represented only one aspect of the man. It was a selective reading, a representation consciously designed to coincide with certain currents in modern environmental thought. When present-day environmentalists claim Wordsworth or any other historical figure as an authority for their views, we must examine this as critically as any other truth claim (see Chapter 2).

Coates (1998) also cautioned against claiming the Romantics for the cause of modern environmentalism. He asked the question: 'Do the hordes of contemporary visitors who come close to overwhelming many national parks on summer weekends experience a spiritual euphoria comparable to that of the nineteenth-century nature poets?' He also argued against overstressing the 'back-to-nature' enthusiasms of the original Romantic Movement. Few wanted to trade permanently the benefits of modern life for the charms of existence in rude nature. Most sought only a temporary antidote. 'Cataracts and mountains are good occasional society', Wordsworth conceded, 'but they will not do for constant companions'.

Anti-industrialism

Later nineteenth-century currents of Romanticism adopted an explicitly anti-industrial stance. Like other cultural dynamics from the Romantic Age, this has influenced thinking about nature, landscape and environment down to the present day. The works of John Ruskin (1819–1900) and William Morris (1834–96) in Britain and Emerson and Thoreau in the USA (see above, p. 143) are frequently discussed in terms of their relevance to present-day environmentalism. As in the case of Wordsworth, there were grounds for co-opting their names to the environmentalist critique of modern industrialism. In Clarke's words (1993, 131): 'This period was in part a re-enactment of an earlier engagement found by the Romantics against the Enlightenment, and we can see in the writings of Englishmen such as Ruskin and Morris, and Americans such as Emerson and

Thoreau, a determination to hold the burgeoning materialism at bay and to re-endow nature with value, spirit, and even divinity.'

Like the earlier Romantics, these writers frequently looked back to, and even sought to recapture, an earlier age when they believed that humanity was in environmental and spiritual harmony with nature. In this regard, they foreshadowed some aspects of thought within today's environmentalism. The writer and critic John Ruskin came from a deeply religious evangelical background that rejected the authority of the established Church. At Oxford, he came under the influence of the geologist the Reverend William Buckland, who taught that geology could be used to prove the truth of the Bible. Ruskin collected fossils and geological specimens and built this fusion of Biblical and scientific truth into his own view of nature. Peter Fuller (1989, 187) called this perspective 'Natural theology'. For Ruskin, the natural world was seen, quite literally as an expressive work of art produced through the supreme imaginative and creative capacities of the Divine Creator. Thus he believed that nature was the model of artistic creativity. In one famous passage in *Modern Painters*, Ruskin imagined the Alps as a 'great plain with its infinite treasures of natural beauty, and happy human life, gathered up in God's hands from one edge of the horizon to the other, like a woven garment, and shaken in deep folds, as the robes droop from a king's shoulders' (quoted in Fuller, 1988, 17–18).

In turn, Ruskin was a formative influence on the group of artists known as the Pre-Raphaelite Brotherhood. William Morris, an associate of the group, became influential in turning Ruskin's criticism of nineteenth-century aesthetic and moral standards into a political and social critique involving a wholesale rejection of the social and economic standards of the day. Coates (1998, 155–6) called William Morris the linchpin of the 'back to nature movement' notably through his novel, *News From Nowhere: or an epoch of Rest* (1891). In this book, Morris's hero fell asleep in his house on the banks of the Thames at Chiswick in the nineteenth century. On awakening in the twenty-first century, he was astounded to find that the vile, stinking soapworks have disappeared, and was delighted that all industrial sights and sounds had been banished. The industrial state had been superseded by an economy combining small-scale, low technology industrial pursuits with agricultural work. This vision combining local and small-scale production appeals to many modern environmentalists, but, as in the case of Wordsworth, it is dangerous simply to co-opt Morris for modern purposes. Though Pepper (1996, 215) believed that the book is important as a critique of industrial capitalism, he conceded that it tends to reproduce a common failing by imagining a golden age of rural harmony and tranquillity drawing on currents in Western thought dating back to the Classical pastoral (see Chapter 4). Its idealised Medievalism endowed the countryside and the past with harmonious social relations. An influential member of the Social Democratic Federation, whose politics developed towards socialism under his guidance, Morris both idealised the Medieval artisan and avoided the problem of how to improve the environment for the mass of the population (see also the discussion of utopianism in Chapter 8).

The activities of later nineteenth-century Romantic anti-industrialists were not confined to the realms of philosophical theorising or artistic representation. Utopian socialism influenced a range of small-scale experimental communities, as well as movements encouraging healthy outdoor recreational activities such as walking, camping and cycling that became particularly influential for attitudes towards nature in the period between 1918 and 1939 (Matless, 1998). For example, Edward Carpenter (1844–1929), poet and founder of the Sheffield Socialist Society, was greatly enamoured of Thoreau's *Walden* and practised self-sufficiency on a smallholding. He advocated return to a state of nature as a cure for the 'disease' of civilisation. In the 1920s, these ideas were put into practice in the alternative scouting group the Kibbo Kift Kin and its offshoot the Woodcraft Folk (Coates, 1998, 156).

Exercise 5.5

The following passage comes from the essay 'Ugliness' in *Nottingham and the Mining Country* (1929) by D.H. Lawrence (1885–1930). In the extract that follows, Lawrence reflected on the town in which he was born and the impact of industrialisation on the countryside and the lives of the inhabitants.

Read through the extract, then briefly itemise the ways in which Lawrence expresses typically Romantic perspective on the relationships between humans and nature.

How do Lawrence's ideas move beyond those of conventional Romanticism?

The collier fled out of the house as soon as he could, away from the nagging materialism of the woman. With the women it was always: This is broken, now you've got to mend it! Or else: We want this, that, and the other, and where is the money coming from? The collier didn't know and didn't care very deeply – his life was otherwise. So he escaped. He roved the countryside with his dog, prowling for a rabbit, for nests, for mushrooms, anything. He loved the countryside, just the indiscriminating feel of it. Or he loved just to sit on his heels and watch – anything or nothing. He was not intellectually interested. Life for him did not consist in facts, but in a flow. Very often, he loved his garden. And very often he had a genuine love of the beauty of flowers. I have known it often and often in colliers. . . .

The real tragedy of England, as I see it is the tragedy of ugliness. The country is so lovely: the man-made England is so vile. I know that the ordinary collier, when I was a boy, had a peculiar sense of beauty, coming from his intuitive and instinctive consciousness., which was awakened down pit. And the fact that he met with just cold ugliness and raw materialism when he came up into daylight, and particularly when he came to the Square or the Breach, and to his own table, killed something in him, and in a sense spoiled him as a man. . . .

Now though perhaps nobody knew it, it was ugliness which betrayed the spirit of man, in the nineteenth century. The great crime which the moneyed classes and promoters of industry committed in the palmy Victorian days was the condemning of the workers to ugliness, ugliness, ugliness: meanness and formless and ugly surroundings, ugly ideals, ugly religion, ugly hope, ugly love, ugly clothes, ugly furniture, ugly houses, ugly relationship between workers and employers. The human soul needs actual beauty more than bread. The middle classes jeer at the colliers for buying pianos – but what is the piano, often as not, but a blind reaching out for beauty?

As a matter of fact, till 1800 the English people were strictly a rural people – very rural. England has had towns for centuries but they have never been real towns, only clusters of village streets. Never the real *urbs*. The English character has failed to develop the real *urban* side of a man, the civic side. Siena is a bit of a place, but it is a real city, with citizens intimately connected with the city. Nottingham is a vast place sprawling towards a million, and it is nothing more than an amorphous agglomeration. There *is* no Nottingham, in the sense that there is Siena. . . .

That silly little individualism of 'the Englishman's home is his castle' and 'my own little home' is out of date. It would work almost up to 1800, when every Englishman was still a villager, and a cottager. But the industrial system has brought a great change. The Englishman still likes to think of himself as a 'cottager' – 'my home, my garden'. But is puerile. Even the farm labourer to-day is psychologically a town-bird. The English are town-birds through and through, to-day, as the inevitable result of their complete industrialization. Yet they don't know how to build a city, how to think of one, or how to live in one. They are all suburban, pseudo-cottagy, and not one of them knows how to be truly urban.

Lawrence was born and brought up in the coal-mining village of Eastwood on the Nottinghamshire–Derbyshire border. His father was a miner and his mother a schoolteacher. He won a scholarship to Nottingham High School, and after university he also became a teacher. He eloped with a married woman and lived abroad for most of his life. He set much of his fiction in the countryside close to his home mining village and tensions between his practical father and aspiring mother were a recurring theme in his work. For Lawrence nature was the spiritual energy of life, producing basic drives and instincts that should be valued as pure truths unclouded by human culture and society. This form of late-Romantic 'vitalism' is so called because it made the core of nature the 'vital spark' of being and consciousness. It had its origins in late nineteenth-century philosophy and became very influential through to the middle of the twentieth century.

Though writing nearly a century after the end of the 'Romantic Age', these passages exhibited many of the characteristics of a Romantic perspective on the relationship between humans and nature:

- The miner was celebrated as a source of real truths about nature, since he lived in intimate connection with the physical world above and below ground.
- True knowledge of nature was based on close observation and the experience of nature as an intimate flow of perceptions, making words and 'civilised' forms of representing nature wholly inadequate.
- Women were represented as symbolic of a false, acquisitive materialism; this privileging of male experience was typical of a Romantic mythology of the masculine hero and heroic suffering artist. It also suggested some parallels with the transformation of the idea of 'Mother Nature' in the Renaissance and further showed the extent to which Romanticism depended on the Enlightenment thinking that it tried to oppose.
- Miners had an intuitive grasp of beauty and crave beauty in fulfilment of lives that are impoverished by the sham and ugliness of modern capitalism.
- Industrial life had robbed honest labourers of fulfilment in their lives and only new forms of social and economic organisation could restore beauty and therefore make their lives complete.

Yet as can be understood from the final two paragraphs of the piece, Lawrence was doing more than simply arguing either for a lost rural idyll or that people should refresh themselves in weekend escapism. Living in an area where ribbon developments of suburban housing and weekend trips into the Derbyshire Peak District were a reality in the years following the beginning of the twentieth century, Lawrence asked the English to face up to the fact that they lived in an urban society (see Chapter 7). The key to his thinking perhaps lay in the term 'pseudo-cottagy', with Lawrence conceivably arguing that all things should be 'true to their natures'. This sort of essentialist thinking was compatible with Romanticism and has roots in classical thought. At the same time, Lawrence pointed the way forward to modernist aesthetics, which value nature as a dispassionate and 'objective truth' without the trappings of religious or other mystical symbolism.

Conclusion

It is important to understand Enlightenment and Romantic approaches to nature both as part of the broad development of thinking about environmental representations and because modern environmental values and actions relate directly to this legacy. Present-day environmental representations show the extent to which these ideas have become central parts of the taken-for-granted ways in which people think about themselves in relation to the world. As a final reflection on these ideas, we would ask you to consider Exercise 5.6.

Exercise 5.6

The two car advertisements in Figures 5.11 and 5.12 portray highly contrasting representations of the environment and society's relationship to it. Figure 5.11 is an advertisement for the British brand leader in the market Land Rover and Figure 5.12 is for its American equivalent the Jeep. Both are luxury cars aimed at an urban middle-income market, rather than off-road working vehicles to be used solely by farmers.

Examine the structure and content of each image.

- How is nature represented as landscape in each?
- What is the relationship between humans and the landscape in each?
- How does each relate to Enlightenment and Romantic conceptions of society–environment relationships?

There are clues in the layout of the landscape, figures and vehicles as well as the supporting text.

When considering the role of these vehicles in contemporary society, some general thoughts might include:

- Urban societies like to maintain some sort of a rural myth that connects industrialised society with nature. Ownership of a four-wheel drive vehicle symbolises this imagined connection.
- Four-wheel drive leisure vehicles are expensive to run, relatively resource inefficient and intrusive in fragile rural environments. Their use reveals an inherent contradiction in Western attitudes to nature that claims to care while selfishly wasting resources.
- Nature is a fashionable commodity, living or spending time outside the city is a leisure and lifestyle choice enabled by personal discretion and available spending power.

Figure 5.11 *Advertisement for Land Rover. Courtesy of Land Rover UK*

Figure 5.12 *Advertisement for Jeep. Courtesy of DaimlerChrysler UK*

Table 5.3 *The characteristics of the Land Rover and Jeep advertisements*

Land Rover	Jeep
High viewpoint	Low viewpoint
Hilly landscape but evidence this is managed agricultural countryside	Little evidence of agriculture suggests wild untamed countryside
Figures seated in car looking down on the landscape	Figures very small in boat on lake
Humans commanding the landscape	Humans dwarfed by wild nature
Text suggests possession, 'all this could be yours'	Text suggests getting away from civilisation and close to nature, 'missing presumed lost'

Turning specifically to the advertisements, your answer could include the elements shown in Table 5.3. The Land Rover advertisement referred back to ways of representing the landscape that relate to the country house prospect (see Figure 5.2). The high viewpoint and the text together suggested ownership. The distant views over a farmed countryside represented an ordered, controlled and agricultural landscape. Here the advertisement made a connection between the Land Rover as a vehicle driven by the British rural farming and landowning communities and the idea that its go-anywhere capabilities would allow town dwellers access to landscapes and scenery at the weekend otherwise unavailable to them. This, in turn, played on the British ruralist myth in which many aspire to both a place in the country and the lifestyle of a country squire without the responsibilities and difficulties of having to make a living off the land.

By contrast, the Jeep advertisement placed the owner of the vehicle in a wild landscape far from signs of human habitation and control. The figures were set small and low down in the image. As such, they appeared humble and insignificant in the face of nature, characteristic of Romantic views of wild landscapes and adopting conventions of the sublime (see Figure 5.9). The figures in the boat were men on a fishing trip – fishing and other forms of outdoor pursuit like hunting, camping and walking were central to the high value placed on intimate knowledge of nature and its restorative powers derived from Romantic thought. Like Figure 5.11, this gave the advertisement a historical reference; in this case, it is to American ideals of wilderness as nature to be tamed and the related ideology of escape from civilisation as a source of masculine identity. These notions were central to European concepts of America as a self-made country where the values of independence and self-reliance are paramount. Here Enlightenment and

Romantic thought together show their importance to Western practices of imperial expansion, control and exploitation.

Further reading

On British landscape since the Renaissance, see:

Michael Reed (1990) *The Landscape of Britain: from the beginnings to 1914*, London: Routledge.
Ian D. Whyte (2002) *Landscape and History since 1500*, London: Reaktion Books.

On the history of landscape gardening, see:

Tom H.D. Turner (1986) *English Garden Design: history and styles since 1650*, Woodbridge: Antique Collectors' Club.
Tom Williamson (1995) *Polite Landscapes: gardens and society in eighteenth-century England*, Stroud: Alan Sutton.

On the history of tourism, see:

Ian Ousby (1990) *The Englishman's England: taste, travel and the rise of tourism*, Cambridge: Cambridge University Press.
John Urry (1990) *The Tourist Gaze: leisure and travel in contemporary societies*, London: Sage.
John Urry (1995) *Consuming Places*, London: Routledge.

6 Empire, exploitation and control

In this chapter, we consider:

- imperialist approaches to nature and landscape;
- key themes with continuing relevance to present-day environmental values.

Like earlier periods, the age of imperialism had distinctive approaches to nature and landscape. Typical ideas about nature and landscape that we highlight in this chapter, include:

- a systematic basis for knowledge of nature;
- approaches to nature and culture informed by Darwinian ideas of survival of the fittest;
- scientific justification of European supremacy as 'natural';
- knowledge of nature formalised in institutions linked to government and military;
- knowledge of nature that enables the control and exploitation of resources;
- progress and improvement becoming a moral, religious and cultural crusade.

Evolution, race and empire

The development of European trading and colonial ambitions in Asia, Africa, the Americas and the Pacific, epitomised by Captain James Cook's three Pacific voyages between 1768 and 1779, encouraged new responses to nature in fields as diverse as painting, horticulture, philosophy and ethnography. The publication of Montesquieu's *The Spirit of Laws* in 1748 raised interest in the impact of environmental forces on society and the environmental changes that resulted from human action (Secondat, 1750). With the revival of Hippocratic thought (see Chapter 4), medical discourse increasingly centred on the interdependence of climate, topography and health. Many diseases were attributed to the effects of 'miasmas' resulting from swamps, marshes and damp ground. The new enthusiasms for health spa resorts, sea breezes and coastal resorts further testified to intense preoccupation with both the malign and beneficial effects of the environment (Arnold, 1996, 20). Montesquieu's ideas about the environment became central for political economy, philosophy, history and geography, with echoes in works such as Adam Smith's *Wealth of Nations* and Hegel's *Philosophy of History* (Arnold, 1996, 24). H.T. Buckle's *History of Civilisation in England* (1857–61), for example, contrasted European and Asian civilisations. Civilisation in Asia, he believed, had natural advantages, particularly the abundant fertile soil of its vast river basins and deltas. Europe was less favourably endowed, but the skill and energy of its inhabitants had overcome its relative disadvantages. Consequently, Europe's growing sense of superiority and strength relative to Asia and the rest of the world reflected its unique ability to surmount and subordinate the forces of nature (Arnold, 1996, 25).

Representations of the environment as the producer of 'natural' social structures and relations could justify inequalities and exploitations in a wide range of contexts. Thomas Malthus' *Essay on Population* (1798), for example, argued that the population was growing too quickly and would exhaust finite food resources. Uncontrollable growth would wipe out the replenishing cycles. Catastrophes such as plague and famine were not aberrations, therefore, but an important part of nature's design. Famine provided the checks and balances that stopped the population outrunning the resources available. As Mick Gold (1984, 25) observed:

> The dinner table, the gardening shears, and the audit book: all these images of a prudent, commercial society were used to legitimise a profound callousness towards 'natural' catastrophes. Naturalists had begun to analyse and classify the animal kingdom; soon this spirit of detached observation gave rise to a similar look at human society.

The inspiration for much of this new mood of analysis came from the work of Charles Darwin's *The Origin of Species* (1859). Darwin's account of evolution provoked remarkable and lasting controversy through its portrayal of the relationship between humans and nature. Representing nature as dispassionately increasing its own vitality by weeding out the weakest strains endowed the concept with an aggressive character. It was an idea that also resonated with the experience of industrialising Britain. The metaphor of nature as 'red in tooth and claw' seemed to explain both the ruthless exploitation of workers and the extent to which Britain and the other world powers busily seized and exploited less advanced countries and cultures. Darwin's imagery and argument provided the perfect rationalisation of this process. 'Survival of the fittest' had initially referred to 'the best adapted to the environment'. Out of a series of chance mutations, those creatures that successfully occupied an ecological niche flourished, while others died out. Soon 'the fittest' meant the most powerful and the most ruthless (Gold, 1984, 25–6).

Darwin's argument influenced sociological thought, in the shape of so-called Social Darwinism, and featured in everyday figures of speech such as 'the rat race', 'the pecking order' and 'the law of the jungle'. Although the laws of nature had been interpreted as fixed and unchanging since the time of Newton, this had changed dramatically in light of evolutionary theory. Now those laws involved change and transformation rather than stability and certainty. Environmental determinism combined persuasively with the ideas of natural selection to provide a new focus on race as a way of explaining the history of human–environment relationships. Europe's growing military and economic ascendancy, for example, might be taken as a sign that its peoples were also racially superior, especially when their arrival in many parts of the world was followed by the precipitous decline and even extinction of indigenous peoples.

In a prominent statement published in 1864, the naturalist Alfred Russel Wallace, who had collaborated with Darwin, injected a belligerent new conception of race into the environmental underpinnings of human evolution. Wallace argued that the Earth may once have been inhabited by a single homogeneous human race but, as humans spread out across the globe, they came under the influence of different environments that had left their imprints in racial differences. Harsh soils and inclement seasons, Wallace claimed, stimulated the Europeans' evolution into a hardier, more active and inventive race. Wallace argued that the 'great law' of the 'preservation of favoured races in the struggle for life' would inevitably lead to the extinction of 'all those low and mentally undeveloped populations with which Europeans come in contact'. Noting the rapid demographic declines among indigenous peoples following the arrival of Europeans, he alleged that the North American Indians, the Indians of Brazil, the Australian and Tasmanian Aborigines and the Maori of New Zealand 'die out, not from any one special cause, but from the inevitable effects of an unequal mental and physical struggle. The intellectual and moral, as well as the physical qualities of the European are superior' (quoted in Arnold, 1996, 28). It was a theme that Western theorists worked and reworked as they structured arguments that justified patterns of domination biologically as well as morally.

They applied their theories to a world that was mostly mapped – by the mid-nineteenth-century only the polar regions remained substantially unexplored. Figure 6.1 shows an example of a new 'thematic' type of map that was produced. Thematic maps illustrated not just the physical features of the world, but also the distribution of specific phenomena such as meteorology, human population, languages, volcanoes and commercial activity. They were made possible by the large quantities of scientific and human research data resulting from Western exploration and trade. The most thorough and wide-ranging versions, like that shown in Figure 6.1, came from the German publisher Justus Perthes (Whitfield, 1994, 120). This map was compiled by Heinrich Berghaus, a Professor of Applied Mathematics in Berlin and associate of the geographer Alexander von Humboldt. Berghaus aimed to depict cartographically all the important aspects relating to humans and their environment. His atlas consisted of some 80 maps, within the major fields of climate, hydrology, geology, natural history and anthropology.

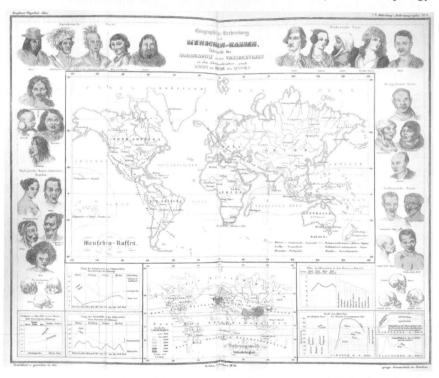

Figure 6.1 *Map by Heinrich Berghaus (1852). Courtesy of the British Library*

Some important points emerge from analysis of the image shown in Figure 6.1:

● The map appeared highly scientific, including a range of graphical and tabular information as well as images of different 'races'. Its scientific rigour, however, was largely illusory. Much of the data on the central map that shows the distribution of races was highly speculative and was neither confirmed nor denied by the supporting graphs showing population density, type of diet, seasons of births and deaths and life expectancy. Rather the scientifically factual graphs provided a *rhetoric* of authenticity (see Chapter 3) for the imaginative distribution of supposed 'races'.

- Berghaus made no overt moral distinctions between the races as such, but that this was a Eurocentric view of comparative strengths and weaknesses was made explicit in one of his later maps where 'mankind's spiritual development' was charted through a system of shading. White areas in Europe indicated the highest state of civilised human life, shading through varying degrees of grey to total darkness for the unenlightened heathen regions of the world. The implications of his link between the study of ethnology and that of moral development were all too clear.

Colonialism and imperialism

From the 1400s to the present, geographical knowledge – the knowledge of places and peoples and the ability to locate them through maps and mapmaking – has been an important tool in the expansion and influence of European countries in distant parts of the world. 'Colonialism' may be defined as the policy of maintaining colonies, the policy of exploitation of 'backward' or 'weak' peoples. The related term 'imperialism' may be defined as the belief in, or desirability of acquiring colonies and dependencies, or extending a country's influence through trade, diplomacy and other means for the purposes of warlike defence and internal commerce.

The age of exploration that lasted from the sixteenth through to the eighteenth century was typified by relatively small-scale colonies and trading posts overseas. These were superseded in the nineteenth century by an age of formal empire, when European countries took formal possession of nations, kingdoms and even entire continents in the Americas, Asia and Africa. At each stage of European expansionism, different forms of empire implied changing ideas of what was nature and what was natural. Worster (1977, 472) characterised imperialism as:

> The view that man's [*sic*] proper role on earth is to extend his power over nature as far as possible. The implicit analogy is with the establishment of one nation's rule over others outside its borders – that is, with the founding of political empires. Francis Bacon (see Chapter 4) was among the first to suggest that a similar dominion should be achieved by the human species over the natural world, through the aid of science-based technology. That science can so serve as an imperialist force has been a recurrent theme in the modern era. It is, in fact, an ethic that has often justified and even directed the pursuit of knowledge.

Smith and Godlewska (1994, 1) commented that the connection between geography and empire is starkly evident. Imperial conquest – whether in ancient Rome, dynastic China or capitalist Britain – invariably involved the geographical expansion of states into other territories: the extent of their territorial acquisition was a rough-and-ready measure of their global power. They quoted the traveller Joseph Conrad in his 1924 essay on 'Geography and Some Explorers' as recognising the relationship between geography, empire and environmental representations by dividing the history of European exploration into three geographical and representational phases:

1   The 'Geographically Fabulous' phase (*c.* 1400–1600), marked by speculative mixes of pre-scientific magic and mythology with graphic and verbal representations of new worlds.
2   The 'Geographically Militant' phase (*c.* 1600–1850), led by a more practical exploration and conquest of foreign seas and territories, exotic species and resources.
3   The 'Geographically Triumphant' phase (*c.* 1850–1940), typified by grand displays of imperial power and ceremony and by explicit theoretical justifications of European domination.

Knowledge and control

As seen in Chapter 5 (Figure 5.2), visual representation is frequently connected to issues of ownership and management. Effective information regarding physical topography was required, of course, in order for a controlling power many thousands of miles away to make strategic decisions. However, the extent to which visual representation was bound into Western belief systems as *the principal source of factual evidence* was itself believed by many to be a product of the new scientific attitude of 'seeing is believing' that dominated during the Enlightenment and still influences the way that people think today (see Rose, 1992). Before the invention of photography, surveying and map-making, the production of inventories and gazetteers went hand-in-hand with the creation of artistic images. Maps produced by explorers during the sixteenth and seventeenth centuries sometimes had images as decorative additions to the cartographic content of the map, showing distant landscapes and peoples to those back home. When Cook voyaged to the Pacific in 1774, drawings were made of the coastline in order to give some context to the cartographically based coastal survey. On his third voyage to the Pacific, Cook chose the artist John Webber (1752–98) to accompany him in order 'to preserve, and to bring home' images of 'the most memorable scenes of our transactions' (Burke, 2001, 130). Figure 6.2 shows one of the charts that Webber drew.

Figure 6.2 *Chart with inset view drawn on Cook's ship* Resolution *in 1774. Courtesy of the British Library*

Other changes to cartographic conventions occurred back in Britain. The accomplished English landscape artist Paul Sandby (1731–1809), for instance, also worked as a cartographer and surveyor for the Ordnance Office in London. Sandby became the most renowned topographical painter of the eighteenth century and is credited with popularising the sites and scenes of the Highlands of Scotland (Daniels, 1993, 61). After the Battle of Culloden (1746) and defeat for the Jacobite cause, Sandby formed part of the military survey team that mapped Scotland, beginning with the Highlands. The survey was an important tool in the control and suppression of the Highlands by a now English-dominated political system. Amongst its practical consequences, the survey facilitated the road-building programme already begun under General Wade after the earlier Jacobite rising of 1715; roads that enabled troops to be moved around the country enforcing and maintaining civil and political order (Reed, 1990, 270–1).

Exercise 6.1

Figures 6.3 and 6.4 show two versions of the same view. The first was a drawing made by Sandby whilst he was engaged on the military survey of Scotland. The second was a picturesque reworking of the view for an English based civilian middle-class audience.

Make a list of the differences between these two images.

Your list could include the following:

- Picturesque coulisse on the left with gnarled trees replacing straight ones.
- Inclusion of rustic figure tending sheep in the foreground (like a painting by Claude).
- Inclusion of a rough rock strewn foreground.
- More elevated perspective for the viewer.
- Distant mountains made to look higher, more jagged and romantic.
- Reflections in the water emphasised to give the picture an element of 'beauty' to balance the roughness of the trees and the foreground and the 'sublime' grandeur of the mountains.

Figure 6.3 View of Strathtay (1747), pen drawing by Paul Sandby (1730–1809). Courtesy of National Gallery of Wales, Cardiff

Figure 6.4 View of Strathtay *(1780), engraving from Sandby in 150* Select Views in England, Wales, Scotland and Ireland. *Courtesy of the British Library*

As this example shows, imperialism and colonial control extended European landscape conventions and ways of seeing landscape into other parts of the world. Just as environments close to home were quickly assimilated into the mode of seeing and value systems of dominant groups, then exotic environments like the Pacific islands were also opened to interpretation using the conventions of European landscape aesthetics. Early European visitors presented Tahiti as a Claudian style paradise and New Zealand as a romantic wilderness complete with exotic 'banditti' (Maoris). Banditti were romanticised images of rogues, robbers and highwaymen that peopled the pictures of Salvator Rosa. In Australia, as Paul Carter (1987) showed, the picturesque, and trees in particular, became highly valued elements in the landscape as settlers tried to come to terms with new landscapes and impose a familiar set of values on new environments. Similarly Europeans re-imagined the Kandyan Highlands of Ceylon by using picturesque, romantic and utilitarian aesthetics, sometimes in combination (Duncan, 1999). This example from L. De Butts' *Rambles in Ceylon* (1841, 161; cited in Duncan, 1999, 159) combines picturesque and utilitarian perspectives:

> The plains [around Kandy] comprise a vast extent of beautifully undulating
> country, dotted here and there with groups of large and majestic trees. The

intervals between which are open and entirely free from jungle. The whole bears a striking resemblance to an English park on a large scale, which would be complete but for the total absence of cultivation and the dwellings of man. A deathlike stillness seems to reign over this apparently deserted valley, and contrasts strongly with the busy, animated aspect of waving corn fields and happy hamlets that adorn the smiling face of an English countryside.

Figure 6.5 shows how the environment of the African savannah was represented to Europeans in complex images drawing carefully on both familiar conventions and unfamiliar images. By doing so, Africa was ultimately transformed into the familiar terms of the aristocratic leisure pursuit of hunting. It is taken from *Travels into the Interior Parts of Africa* by François Le Valliant (1790); probably the most widely read eighteenth-century account of travel in the Southern African region and a major best seller in England.

Figure 6.5 Encampment in the Great Namaqua Country. *Etching, frontispiece to* Travels into the Interior Parts of Africa, *by way of the Cape of Good Hope (1790) by François Le Valliant*

Things to note include:

● The scene was staged as in a theatre, which was typical of illustrations in many explorer texts. The vegetation had a framing effect, like a proscenium arch; the figure appeared to be drawing aside a curtain of bushes, leading the viewer into the landscape and the reader into the book.

- The picture showed clear evidence of the use of picturesque conventions. There was a high vantage point flanked by Claudian coulisses (see Chapter 4). The space was clearly divided into raised foreground, the middle distance (which contained a prominent object of interest), and a light-saturated horizon.
- Items in the picture were ordered in terms of familiarity for viewers. At the outer edge of the frame there was a Europeanised tree, leading in to an exotic palm, with the giraffe placed centrally in the middle distance. The giraffe would be the most unfamiliar and exotic aspect of the picture, but equally giraffes had great value for European collectors of exotic species. They were also difficult to represent and classify within the developing system of European biological science (Bunn, 1994, 131).
- The gestures and stance of the European figure in the foreground were copied from a European etiquette book. The pose represented the ideal stance for giving a command. The hunting dogs were a *metonym* (see Chapter 3) of the European explorer's ability to command authority back home and by implication in Africa also. Thus the scene was transformed into a conventional aristocratic hunting scene in a European parkland landscape and the foreign landscape was brought under a form of aristocratic control very familiar to the readers of Le Vaillant's guidebook (see Bunn, 1994, 129–32).

Hunting, as a sport and as a metaphor, played an important part in imperial expansion. In many areas of the world, the colonial frontier was also a hunting frontier and animal resources contributed to the expansionist urge. During the period of conquest and settlement, animals formed an important subsistence and economic resource. By the 'high noon of empire' during the late nineteenth and early twentieth centuries, 'hunting had become a ritualised and occasionally spectacular display of white dominance' (MacKenzie, 1988, 7). In the nineteenth century 'great' white hunters and collectors like F.C. Selous, C.H. Stigand, Denis Lyell and Richard Meinertzhagen gained an *entrée* into the aristocratic elite and into the scientific circles of natural history museums through their hunting prowess. Social status and scientific authority combined with the danger of hunting and the technological power to kill 'native' animals. Together these attributes of hunting contributed to justifying imperialism in Social Darwinist terms as the 'survival of the fittest'.

Control of nature was easily extended to the control and subjugation of local indigenous populations. As MacKenzie (1988, 308) remarked, until the 1930s: 'Hunting remained the image of war, calling forth qualities of courage, judgement, knowledge of terrain and the ability to inflict violence and death. Yet it was an old-style war that was evoked, a war of rules and "sporting" attributes.' Overwhelming power plus civilised behaviour together formed cornerstones for the representation of imperialism. In Britain it was argued that hunting produced a heroic, resourceful, fit and healthy breed of young men necessary for national security at home and the advance of civilisation abroad. As such its codes and practices were represented in journals, popular stories, the activities of the Boy Scouts and, above all, the cinema. The idea of Africa as a 'game reserve', a playground for rich Europeans and Americans and the transformation of African mammals – 'big game' – into the trophies of conquest had an enduring influence on European attitudes to African environments. The reinvention of hunting as the 'viewing safari', based around the camera and the photograph rather than the rifle and the trophy, retains the essential element of the hunt. Though its influences on African mega-fauna are more benign and nowadays tied into eco-tourism and conservation strategies, the wildlife safari still engages with the Western fascination with the control and possession of nature. Whether imagined as 'game reserve' or 'national park', the dominant representation of Africa is still that of free and uncharted wilderness available for Westerners to take and deploy as they wish.

By the end of the nineteenth century, European nations commanded almost all of Africa and vast areas of Asia. The imperialist perspective on nature shifted from the state of all-out war that characterised the 'Darwinian' discourse to the need for efficient management. Having conquered foreign territories, the task was to establish efficient government. Environmental ideas were refashioned to meet the ideological imperatives of a new imperial age. Naturalists, anthropologists, historians and geographers were actively involved, spurred on by an emerging sense of professionalism and impelled by involvement in empire as a moral crusade. The combination of racial Darwinism with an ascendant Western imperialism placed environmental ideas into a position of exceptional influence and prominence between the 1890s and the late 1920s. As Mick Gold (1984, 26) put it: 'Having eliminated your industrial competitors, the job was now to consolidate your market. Having collected all the animals and plants, it was now necessary to catalogue them.'

Major museums systematically categorised their collections, giving order to the vast range of specimens killed and brought back to Europe. In Britain, Lord Rothschild despatched expeditions to remote corners of the globe to collect butterflies. The Rothschild Collection ultimately amounted to 2.5 million set butterflies and moths, meticulously labelled and mounted, alongside 300,000 birds, 144 giant tortoises and 200,000 birds' eggs. The Royal Botanic Gardens at Kew (London) housed a similar collection of plants from all over the world. Research and selective breeding of plants took place at Kew. Important crops were selectively bred and then transplanted to other locations within the British Empire. The rubber plant is a good example. Successfully grown from seeds smuggled out of Amazonian Brazil in an umbrella and then improved upon, it was transplanted to Malaysia where it became the basis of a vast industry.

In Oxford, the Pitt Rivers Museum (Figure 6.6) was founded in 1884 as an anthropological museum, with curatorial policies strongly influenced by Darwinism. Pitt Rivers saw cultural evolution as being like biological evolution and held the view that humanity moved through stages of 'progress' towards civilisation. He believed that objects should be classified like biological species and coined the word 'typology' to describe the progressive development of specific aspects of human culture. Pitt Rivers was convinced that anthropological science could trace a 'predictable course of gradual improvements' in the evolution of any material object – whether a house, a hammer or a musical instrument – from its earliest beginnings to its most sophisticated form. Hence, the exhibition of ethnographic artefacts in the Pitt Rivers Museum is based on typological, rather than geographical or cultural divisions, thereby emphasising their development through evolution.

Figure 6.6 *The Pitt Rivers Museum, Oxford*

The Pitt Rivers Museum displays culture as an imperialist counterpart to Kew Gardens or the Natural History Museum in London. Its location is significant, since Oxford was an important fulcrum of empire. Within the immediate vicinity are: the University Museum, the site for the famous debate concerning Darwinism between the scientist T.H. Huxley and Bishop Wilberforce; Rhodes House built to accommodate the bequest of Cecil Rhodes (1853–1902), 'founder' of Rhodesia (now Zimbabwe); the former Imperial Forestry Institute; and the Indian Institute. University graduates dominated political and diplomatic service, both at home and in the empire. The Pitt Rivers Museum itself played an important role in the latter. It educated men entering the Colonial Service about the customs and cultures of those peoples whom they were going to manage. It is perhaps not surprising then that many of the major collectors for the museum were in the Colonial Service themselves.

Figure 6.7 shows a map of the British Empire drawn by Walter Crane in 1886. It was perhaps 'less a map than an icon, for rarely can a map have expressed a political philosophy so clearly . . . Walter Crane created here a cavalcade of imperial subjects' Whitfield (1994, 124). The moving force behind this map was not war, but peace and good government. Crane chose as symbolic elements in his composition: the labour of the farmer and the huntsman; the fruits of earth and sea; exotic wildlife; the beauty of the noble savage; and the graces of the east. The whole tableau unfolded beneath the gaze of Britannia, her Greek helmet recalling Minerva's twin gifts of strength and beauty. While the doves of peace fluttered in the map's margin, her rule was clearly guaranteed by force of arms on land and sea.

Figure 6.7 *The British Empire (1886) by Walter Crane*

The map was edited by Sir John Colomb, a prolific author on imperial and military subjects, and was published in *The Graphic* magazine. The small inset map contrasted the extent of British territories in 1786 with the situation a hundred years later. As such:

> This map clearly functions on several levels – those of geography, art and politics – and so places the world map firmly in the realm of popular culture. The psychological influence of such maps can only be guessed at, but it is impossible not to believe that the elements of the whole – the map, the imagery, the concept of Empire – did not become merged in the English mind into a mesmeric image of the world and its people.
>
> (Whitfield, 1994, 124)

Exercise 6.2

Compare this map (Figure 6.7) with the Psalter Map (Figure 4.4), the World Map by Pieter van den Keere (Figure 4.7) and the map by Heinrich Berghaus (Figure 6.1).

Can you fit them into the threefold typology of imperial representations suggested by Smith and Godlewska (see above, p. 160)?

Your list might include:

- The Psalter Map (*c.* 1250), shown in Figure 4.4, fits comfortably into Smith and Godlewska's characterisation of the 'Geographically Fabulous'. It was an image firmly based in pre-scientific magical and mythological representations of distant lands. It included Biblical events such as Noah's Ark, the crossing of the Red Sea and the walls imprisoning Gog and Magog, along with fantastical representations of African peoples as the 'monstrous races'.
- The map by van den Keere (1611), shown in Figure 4.7, reveals that knowledge of the world was rapidly increasing through exploration and trade, but the image was also symbolically rich. Its dominant themes were power, kingship and international treaties. The map therefore reflected the themes of the 'Geographically Militant' – practical exploration and conquest – although it also took the form of a highly pictorial symbolic representation.
- The map by Berghaus (1852), depicted in Figure 6.1, purported to show the results of Western scientific knowledge but much of this information was either irrelevant or inaccurate. The map fits into the theme of the 'Geographically Triumphant' because it was part of a theoretically argued justification of European domination. At the same time, the aggressive approach to race that informed this map was part of the concluding phase of militant conquest.

- The British Empire (1886) by Walter Crane, shown in Figure 6.7, was an archetypal image of the 'Geographically Triumphant' theme. Here Empire was settled and secure. Britain was situated amid clear evidence of its imperial success.

Difference, resistance and exploitation

For some authors the history of landscape is almost synonymous with processes by which the wishes of the powerful come to be imposed on the powerless (Rose, 1992, 1993). This perspective, for example, has informed studies of elite land-scaping practices in eighteenth-century England (see Chapter 5), urban landscapes of consumer and financial capital (see Chapter 8; also Domosh, 1996; McDowell, 1997) as well as the activities of European colonists. From the beginnings of European overseas exploration, encounters between colonists and indigenous peoples revealed differing interpretations of the relationships between people and land. The European landscape ethic saw land as a tradable commodity and private property, having value only to the extent that it was a source of expropriable material wealth. This contrasted with an indigenous landscape ethic that prioritised spiritual values, stewardship, communal use and inviolable rights to dwell (Young, 1992; Hirsch and O'Hanlon, 1995; Ellen and Fukui, 1996; Strang, 1997; Bender and Winer, 2001).

Cultural differences over landscape became institutionalised in treaties and agreements, the enforced concentration of colonised peoples into homelands and 'reservations'. Pawson (1992) showed the different cultural contexts that the Maoris and the British brought to the Treaty of Waitangi in 1840. The Treaty consisted of a brief statement of three articles and was prefaced by a preamble in which the Crown offered a partnership to the Maori people. This established a co-operative context that would have been readily appreciated by representatives of a culture structured on group relationships. There were, however, crucial differences between the English and Maori language versions of the Treaty, centred on problems of translating concepts intrinsic to one culture into the language of another. The Maori version signed by the chiefs guaranteed them 'te tino rangatira-tanga', chiefly power over the land. It only ceded 'kwanatanga' or governorship to the British. By contrast, the English version secured the Crown's sovereignty over all New Zealand. As Pawson (1992, 20) observed:

> That which mattered most to the Maori, that which they thought, and many still think, the treaty protected, was the very thing that experience proved was to be taken away. As an influential chief said in 1840 'the shadow of the land goes to the Queen, but the substance remains with us'. Just a year later he revealed his fear that 'the substance of the lands would pass to the European and only the shadow would remain with the Maori people'.

The contested nature of landscape is thus an ongoing source of conflict. Culturally specific conceptions of landscape and the social and material inequalities that these imply are subject to defence and challenge in, for example, North America, South Africa, New Zealand and Australia. Not all Western representations of the colonial landscape, however, were insensitive to the presence of indigenous populations. Figure 6.8 shows a view of the Bay Islands of New Zealand, painted around 1827 by Augustus Earle (1793–1838). Earle was a London-born travel artist. In 1818 he left England on the first stage of a round-the-world journey that took him to South America, Tristan da Cunha, New South Wales, New Zealand, the Pacific, India, Mauritius and St Helena. Earle regarded the Maoris as a complex advanced culture; he admired and copied their art, both the woodcarvings and their elaborate tattoo designs.

Figure 6.8 Distant View of the Bay Islands, New Zeeland *(c. 1827) by Augustus Earle (1793–1838). Courtesy of the Rex Nan Kivell Collection, National Library of Australia*

Exercise 6.3

Study the carved figure on the right-hand side of Figure 6.8.

What role does it serve in this apparently conventional European landscape painting?

Things to note in this picture:

- Earle used the conventions of European picturesque landscape; alternating planes of light and shadow give the landscape depth. The Maori carved figure to the right acted as a side screen for the image.
- However, the Maori figure did not, as in a traditional European painting, provide a dark refuge for the viewer to hide behind, nor did it provide a convenient stand-in for the viewers' gaze within the composition. The function of the figure in Maori culture was to stand guard over taboo territory, to separate the sacred, forbidden landscape from the territory surveyed and traversed by the European traveller and his Maori companions. Thus the figure was an emblem of a non-European way of understanding landscape that stared back into the space of the European viewer.
- The Maori bearer on the left seemed hesitant as he walked, turning to the side to scan the taboo territory while raising his war club slightly to ward off a potential threat. The Maori warrior just ahead of the European traveller seemed to be joining in the Western gaze, looking out towards the opening clouds and horizon. His musket, upright posture, and European garments suggested that the Maori chief was able to make the transition from Maori sense of taboo landscape and sharing in the European appreciation of 'prospect' as an elite way of seeing.
- The intermingling of landscape conventions was also evident in the overall composition as well as its details. The picturesque convention of the serpentine line from foreground to background was cut off in the picture and turned back on itself. In place of the serpentine access route the composition deployed a crescent, canoe shaped hollow, enveloping a procession of equally scaled figures across the shallow surface of the painting. The effect was an oval or circular procession. It advanced up towards the viewer on the left, retreated away on the right, eternally suspended on the canoe-shaped threshold between two landscapes – the depicted European picturesque prospect and the implied (non-representable) taboo space of the Maori guarded by the carved figure.

Mitchell (1994, 24–7) argued that Earle had become close enough to Maori culture to recognise that these landscapes were not simply a passive field for colonisation but a site of contested meanings between European and Maori aspirations, conceptions of landscape and nature. This picture represented a 'subtle resistance to European conventions'.

Fantasy and the exotic

Fantastic representations of unknown lands, and the people who lived there, could be found throughout the history of Western exploration. The ancient Greeks imagined so-called 'monstrous races' (see Figure 4.4) living in faraway places such as India, Ethiopia or Cathay. These races included; the dog-headed people (Cynocephali); those lacking heads (Blemmyae); the one-legged (Sciopods); cannibals (Anthropophagi); Pygmies; and a martial, one-breasted race of women (Amazons). In his *Natural History*, the Roman writer Pliny effectively transmitted these stereotypes to the Middle Ages and beyond. Burke (2001, 127) suggested that the 'monstrous races' may have been invented to illustrate theories of the influence of climate, the assumption that living in places which are too cold or too hot prevented people from being fully human.

As India and Ethiopia became more familiar to Europeans in the fifteenth and sixteenth centuries the stereotypes of the 'monstrous races' relocated to the New World. The River Amazon, for instance, takes its name from the belief that the Amazons lived there. Remote peoples were viewed as morally as well as physically monstrous, as in the case of the cannibals believed to live in Brazil, Central Africa and elsewhere. Usually interpretations have focused on representations of non-Western peoples and their cultures rather than non-European environments. Arnold (1996, 141–2) argued that, during the eighteenth and nineteenth centuries, alien landscapes were frequently imbued with as much importance as peoples or cultures themselves. Landscapes, and other aspects of the physical environment, such as climate and disease, were endowed with great moral significance. As shown earlier, environments were widely believed to have a determining influence upon cultures: savagery, like civilisation, was linked to certain climatic or geographical features. Beyond the privileged shores of Europe, nature dictated culture.

The representation of that which is different or unfamiliar frequently takes one of two forms:

- That which is different can be incorporated with that which is familiar as people try to interpret the new using already known codes and conventions. The picturesque and utilitarian approaches to environments new to Europeans exemplify this strategy, which is called *assimilation* or *homogenisation*.
- The common alternative is consciously or unconsciously stereotyping the unfamiliar as the opposite of one's own culture. Such a stereotype is called an *Other* and the process of its creation as a cultural object is known as *Othering*. Others and the process of Othering are fundamental to the way in which identities are created and maintained.

Orientalism and Otherness

Since the Middle Ages, if not earlier, Middle Eastern lands have played a distinctive role in the formation of Western European identity. Located close to Europe's south-eastern edge around the Mediterranean, Arabic civilisations and the ancient civilisations of the Near East have been both a source of language, culture and religion for the West while simultaneously representing something different, even opposite, to Westerner's ideas of themselves. Particularly since the Romantic Age, the Orient has become celebrated as a source of sexually charged exotic beauty, in addition to being feared for its religious devotion and military strength, and envied for its resources and long-established trading networks.

The literary critic and historian Edward Said highlighted the historical and contemporary implications of this long-standing Western perspective on the Middle East in his book *Orientalism* (1978). According to Said, the West has created a dichotomy, between the reality of the East and the romantic notion of the 'Orient'. The Middle East and Asia are viewed with prejudice and racism. They are represented as backward and unaware of their own history and culture. To fill this void, the West has created a culture and history for them based on its own preconceptions and prejudices. On this framework rests not only the study of the Orient, but also the political imperialism of Europe in the East.

As shown in Chapter 4, associating nature with the feminine has played an important role in Western attitudes to environmental exploitation. The representation of women as the *Other* of men – closer to nature, linked to the biological functions of reproductive fertility and subject to 'animal' passions and instincts – contrasts with representations of the masculine as linked to control, rationality and higher cultural pursuits. The idea of nature as female, for example, was important to the notion of America as a sort of virginal *tabula rasa* (Smith, 1970), on which European settlers had the chance to begin anew, retracing the natural cycle of development. This theme persisted in American history through the writings of Thoreau (see Chapter 5) and Frederick Jackson Turner (1861–1932). It was central to the myth, played out in endless films and books, about the heroic pioneers who carved out America. Merchant (1995, 98), for instance, showed how the language of a 'pulsating sexuality of a laughing, vital Mother earth and virile sun' informed the popular representations of nature for pioneer European settlers in New England.

Figure 6.9 shows a sketch map found in the opening pages of Henry Rider Haggard's best-selling novel *King Solomon's Mines*. In the story, the map led three white Englishmen to the diamond mines of Kukuanaland somewhere in southern Africa. The story told how a Portuguese trader, Jose da Silvestre, drew the original map in 1590 while he was dying of hunger on the 'nipple' of a mountain named Sheba's Breasts. Da Silvestre's map promised to reveal the wealth of Solomon's treasure chamber, but carried with it the obligatory charge of first killing the black 'witch-mother', Gagool (McClintock, 1995, 1).

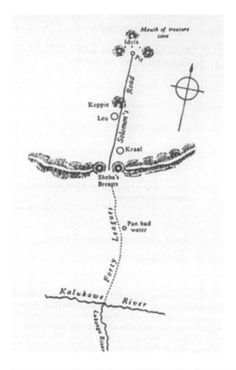

Figure 6.9 *Rider Haggard's sketch map of the route to King Solomon's Mines*

Exercise 6.4

How does the representation of nature as female and 'native' in the map shown in Figure 6.9 justify imperialist exploitation?

Things to note about this image are:

- The map was a conventional treasure map showing a route to the riches of the diamond mines. The heroes of the story had to follow the map and cross this landscape if they were to find the treasure.

- However, if turned upside down the map also symbolically represented a female body. The body is spread-eagled and truncated – the only parts drawn were those that denoted female sexuality. In the story, the travellers crossed the body from the south, beginning near the head, which was represented by the shrunken 'pan bad water'.
- At the centre of the map lay the two mountain peaks as handless arms. Solomon's Road, leading from the breasts over the navel koppie straight to the pubic mound, inscribed the length of the body. In the story, this mound was named the 'Three Witches' and figured by a triangle of three hills covered in 'dark heather'. This dark triangle both pointed to and concealed the entrances to two forbidden passages: the 'mouth of treasure cave' – the vaginal entrance into which the men were led by the black mother, Gagool – and, behind it, the anal pit from which the men would eventually crawl with the diamonds. This male birthing ritual left the black mother, Gagool, lying dead within (McClintock, 1995, 3).

Thus the question that heads discussion of the map may be answered as follows:

- On this imaginative map, the feminine embodied a magical and indigenous form of power contrasting with the European scientific rationality and heroic masculinity of the men. The male quest for treasure, hidden within the feminine body of the foreign earth, reflected and gave imperialist form to the changing conception of 'Mother Nature' in Renaissance Europe (see the discussion of Francis Bacon in Chapter 4). Here the adventurers have to investigate and tame Mother Nature, both as a woman and a 'native', if she were to give up her resources.
- Written as a story for European boys, the map connected success of the story's masculine hero to conquest and riches in sexual, racial and material terms. The hero fulfilling the quest for treasure showed his power by becoming a 'real man' in all senses of the term. The distant lands of empire were 'virgin' lands available for a wide range of fantasies and fulfilments (McClintock, 1995).

Examining this weight of evidence, McClintock (1995, 1–2) remarked:

> In this way, Haggard's map assembles in miniature three of the governing themes of Western imperialism: the transmission of white, male power through control of colonized women; the emergence of a new global order of cultural knowledge; and the imperial command of commodity capital.

Arnold (1996, 142) argued that one of the principal manifestations of environmental Otherness in European thought since the fifteenth century has been an emerging distinction between temperate and tropical lands. The complex of ideas and attitudes known as 'tropicality' represented environmentalism in one of its most influential and enduring forms. Calling part of the world 'the tropics' (or an equivalent term, such as the 'equatorial region' or 'torrid zone') became, over the centuries, a Western way of defining something culturally alien, as well as environmentally distinctive, from Europe and other parts of the temperate zone. The tropics existed only in mental juxtaposition to something else – the perceived normality of the temperate lands (see also Smith, 1985).

Islands supply another important component in the development of Western notions of 'tropicality'. As Tuan (1974, 118) commented, they seemed to exert a tenacious hold on the human imagination, playing an important part in Buddhist, Hindu and Chinese as well as Western culture. Many creation myths begin with the watery chaos: the land on which life has its start appears amid this primeval landscape necessarily as an island. In numerous legends the island appears as the abode of the dead or of immortals. Above all, according to Tuan, islands symbolise the Garden of Eden surrounded by water that protects unspoilt nature from the dangers of civilisation. They are a state of 'prelapsarian innocence and bliss, quarantined by the sea from the ills of the continent'.

In Europe, the legend of the 'Island of the Blessed' first appeared in ancient Greece, described as a place that provided heroes with unusual harvests thrice a year. The Celtic world, although remote from Greece, had a similar legend. The imagination of the Middle Ages peopled the Atlantic with a large number of islands and many of these myths persisted well into the early modern period. European travellers had for centuries set out with their minds filled with images of old or fabled worlds, which they sought to rediscover – Arcadia, Eden, Atlantis, Hesperides, Jerusalem or Cathay. The voyages of Captain Cook largely confirmed the desirability of the South Sea Islands (Tuan, 1974, 119). The image that the Pacific islands overwhelmingly evoked was that of Paradise, with their lush environment, their beneficent climate and their people free of the tyranny, hunger and poverty that beset some European and Near Eastern societies. The character of the Oceanic peoples provoked questions about human nature that chimed precisely with those that philosophers were asking about government, liberty, and happiness in human society. Cook's death in 1779, however, initiated a process of polarisation. Nineteenth-century views divided between those who called the Pacific islands a heavenly paradise and those who viewed them as a region ridden by disease, especially leprosy. Influential books included Herman Melville's *Typee* (1846), a semi-fictional account of his stay in the Marquesas, R.M. Ballentyne's *The Coral Island* (1858) and Harriet Martineau's *Dawn Island* (1838).

The work of the painter Paul Gauguin (1848–1903), who first moved to Tahiti in 1891, was significant in continuing the myth of the Pacific Islands as a tropical paradise. Describing his life in Tahiti he said: 'I have escaped everything that is artificial and conventional. Here I enter into Truth, become one with nature. After the disease of civilisation life in this new world is a return to health' (quoted in Osborne, 1970, 462). Gauguin's bold and colourful paintings of Tahitian women and his own licentious lifestyle in Tahiti seemed to confirm the image of a tropical paradise. Yet Gauguin's paradise presented rather a misleading image. Hughes (1980, 129) put it in terms almost as colourful as Gauguin's paintings:

> The island was far gone. Its decline had begun at the moment Captain Wallis arrived, and had been going on without interruption or help for 125 years. Instead of Paradise, Gauguin found a colony; instead of Noble Savages, prostitutes; instead of the pure children of Arcadia, listless half-breeds – a culture wrecked by missionaries, booze, exploitation, and gonorrhoea, its rituals dead, its memory lost, its population down from forty thousand in Cook's time to six thousand in Gauguin's. 'The natives', he lamented, 'having nothing, nothing at all to do, think of one thing only: drink.'

After several periods of pain and poverty, Gauguin died of syphilis in the house he built for himself in the Marquesas Islands that he called *The House of Pleasure*. Hughes dismissed Gauguin's paradise as 'a defiled Eden full of cultural phantoms'. There seems to be a dark side to even the most positive views of the imperial exotic fantasy. As Arnold (1996, 142) noted: 'Part of the significance of "tropicality" lay in its deep ambivalence. In part an alluring dream of opulence and exuberance – Edenic isles set in sparkling seas – the tropics also signified an alien world of cruelty and disease, oppression and slavery.'

Conclusion

These reflections on fantastic and exotic landscapes draw to a close this survey of the significant phases in the history of environmental representations that are both an important part of cultural history and a persistent and pervasive influence on present-day discourses about landscape and nature. Imperialist representations of the environment reflected changing conceptions of nature across the history of exploration and expansion from the early Renaissance onwards. During the nineteenth century, imperialist approaches to nature drew on both Enlightenment and Romantic practices, synthesising an expanding frontier of scientific knowledge with approaches to human environment relations grounded in Classical thought. Linked to Darwinian evolutionary theory, imperialist perspectives on nature developed racially based theories that justified Western colonial domination. In many parts of the world imperial control was exercised in terms of the contrasting exploitative (individualised) and communal conceptions of land and nature held by Western and indigenous peoples. As shown in the two previous chapters, certain themes recur constantly as we analyse the representational history of Western environments.

Exercise 6.5

Look back to the three themes highlighted at the end of the Chapter 4:

- the pastoral
- animate nature
- landscape and identity.

To what extent were these important to Enlightenment, Romantic and imperialist environmental values and their cultural representation?

The following points relate to this question:

- The pastoral, for example, directly informed elite landscaping during the eighteenth century. In addition, its key theme of a middle ground or bounteous garden landscape situated between barren wilderness and the degenerate city informed both romantic reaction to industrialism and European imaginings of exotic distant lands.
- The idea of animate nature was central to earlier versions of romanticism and important for the transcendentalist approach to nature, which informed the landscape preservation movement in the United States.
- The role of landscape in making and expressing identity is fundamental to the ongoing conflicts described throughout this chapter between the values of landscape as an elite way of seeing and land as a representation of popular and folk values based on working the land. During the period considered here, landscape is also very frequently an expression of national identity. In the Enlightenment it was a metaphor for English democratic civilisation. In Romanticism, it was a manifestation of everyday moral nobility untainted by greed and affectation. In imperialism, it was an expression of legitimate power and control.

We have hinted in this chapter and the two previous chapters at the ways in which these ideas have been transformed and transmitted down to the present, but so far we have left untouched representations of the city – the environment often, if misleadingly, taken to be the inseparable opposite of nature. In the next two chapters we fill this gap as we tackle representations of historic and contemporary urban environments.

Further reading

On the relationship between nature and Empire, see:

David Arnold (1996) *The Problem of Nature: environment, culture and European expansion*, Oxford: Blackwell.

John M. MacKenzie, ed. (1990) *Imperialism and the Natural World*, Manchester: Manchester University Press.

On the idea of wilderness and its particular importance in American thought, see:

Roderick Nash (1982) *Wilderness and the American Mind*, New Haven, CT: Yale University Press.

Max Oelschlaeger (1991) *The Idea of Wilderness*, New Haven, CT: Yale University Press.

For indigenous perspectives on landscape and nature, see:

Barbara Bender and Margot Winer (2001) *Contested Landscapes: movement, exile and place*, Oxford: Berg.

Roy F. Ellen and Katsuyoshi Fukui, eds (1996) *Redefining Nature: ecology, culture and domestication*, Oxford: Berg.

Eric Hirsch and Michael O'Hanlon (1995) *The Anthropology of Landscape: perspectives on place and space*, Oxford: Clarendon Press.

7 Representing urban environments

This chapter, the first of two concerning representations of the urban environment, considers:

- the values, sympathetic and hostile, that have pervaded representations of the city since Classical times;
- reasons for the duality found in representations of the city in Western society;
- the dominance of hostility towards the city since the spread of industrialisation, focusing on the writer's response to mid-nineteenth-century London;
- contrasting representations of neighbourhoods within the American city;
- the dominantly positive, but complex representations of the city found in the art of place promotion, with particular reference to the selling of suburbia and the rebranding of industrial cities.

Good cities, bad cities

The process of representation, as we have shown at various points in this book, is one that builds on culturally derived and endlessly reinforced sets of conventions. Nowhere is this more true than for representations of cities and urban life; subjects about which there are two, sharply polarised discourses. Throughout the ages there have been observers who have believed that city life is inherently beneficial or, at least, contained the conditions for human fulfilment. Equally, there have always been those who believe that the city offers a hostile environment for human life and, quite often, one that imposes severe costs on its residents. The conclusions reached largely depend on the initial values of the observer.

This polarity again has roots in Antiquity. For the philosophers of ancient Greece, the city-state represented the epitome of civilisation: indeed the Greeks 'never produced a political formula which was not deeply rooted in the concept of the *polis*, the politically autonomous city' (Mazzolani, 1970, 16). Plato, for instance, argued that the city constituted the highest expression of human activity, at least potentially. The purpose of politics was to make the city a harmonious living entity governed by reason, which would be achieved by 'intelligent research into what is best for beings living a communal life together' (Pradeau, 1997, 167). Looked at another way, the city became the medium for achieving the ideological goals of its philosopher-rulers. For his successor Aristotle, the city originated in human beings' search for the satisfaction of physical needs, but its justification becomes its ability to satisfy moral needs, the needs of the soul, allowing its citizens to live well. At this point, Aristotle argued that the *polis* was natural and that nature intended human beings to live in the city (Nicholson, 1984, 37).

The Romans assimilated much of Greece's cultural legacy but were more ambivalent about the city. Certainly, Roman writers of all periods never tired of praising the simple, honest life of the farmer and pastorialist, while berating the growing corruption of the city. At the same time, there was a counter-theme of extravagant praise for the city of Rome as the heart and soul of Empire (Vasaly, 1993, 156). It is no coincidence that the Latin word for citizen (*civis*), a person who had the fullest rights under the Roman Empire, shared the same roots as the word for civility (*civilis*). In the Republican period the philosopher and orator Cicero (106–43 BC) depicted the city as the locus of stimulation and sophistication in human affairs. Arguing against those who castigated the luxury and corruption of Rome, Cicero provided alternative description of the city as a determinant of ethics. In his *Pro Caelio*, for example, he used 'the physical setting of the city and the training that could be secured within this setting to imply that – far from being simply a sink of corruption – only contemporary Rome could provide the conditions by which the state might produce for itself the best of men and best of citizens' (Vasaly, 1993, 179). Later, during the pre-Christian Empire, the authorities built temples to celebrate the city's goddess and honour Rome's foundation myths. The underlying reasons for venerating the city differed, but the product served the same purpose. Venerating the city of Rome served both the Empire's spiritual needs and the ideological interests of the ruling classes.

By contrast, the Bible's portrayals of cities, emerging at a similar time, were predominantly negative. Cain, the first builder of a city, committed the first recorded murder. Nimrod, the founder of great cities, was a descendant of the 'cursed line' of Ham. Sodom and Gomorrah, the cities of the Plain, were destroyed by fire and brimstone because of the wickedness of their inhabitants (Genesis 4, 10, 18). The New Testament, dating from the early years of the current era, reflected the history of a people starting to adopt some of the values of Roman culture. Nevertheless, despite being less strongly anti-urban than the Old Testament and culminating in the promise of the sacred city of Revelation 21, it still contained images of the sinful lifestyles of the inhabitants of cities (e.g. Matthew 11). Christian theologians, for their part, developed those ideas further. In *The City of God*, for example, St Augustine (AD 354–430) saw the earthly city poised between Cain and Abel, Babylon and Jerusalem, self and God. The tension between them could not be resolved on Earth, no matter what the social order. Christianity had broken the horizons of the Roman political world by its rejection of the pagan gods. The true believers had to preserve their identity as citizens of heaven and remain resident strangers of the earthly city (Coleman, 1984, 53–4; Dougherty, 1999, 194).

The 'Great Wen'

These dual and polarised discourses co-existed throughout the centuries. London, for example, continually fell in and out of favour with observers, according to their values. At the end of the twelfth century Richard of Devizes castigated its vice and simmering lawlessness: 'Whatever evil or malicious thing that can be found in any part of the world, you will find in that one city.' Yet as late as the early eighteenth century, scholars basked in the city's civilising virtues and enlightenment. Voltaire in the 1730s described London as the 'rival of Athens' and Dr Johnson famously remarked: 'No, Sir, when a man is tired of London, he is tired of life; for there is in London all that life can afford.' Even John Gay, the caustic commentator on human shortcomings, recorded the pleasures as well as the hazards of walking the streets of London in 1716 (Gay, 1716).

Nevertheless, the view from below was causing alarm bells to ring. London's 'gin epidemic' of 1720–70, powerfully caricatured by Hogarth's engravings of 'Beer Street' and 'Gin Lane' (both from 1751), revealed deep cultural anxieties about urbanisation, modernity and mass consumption (Nicholls, 2003). With industrialisation, the pendulum moved sharply and strengthened the discourse of anti-urbanism. In 1783, Josiah Tucker, the Dean of Gloucester, referred to the increase of building in London as 'a Wen, or Excrescence, on the Body Politic' (Tucker, 1783, iii, 45). The Radical MP William Cobbett later reinforced that theme, referring to the same city as the 'Great Wen' in *Rural Rides* (Cobbett, 1821). In 1785, the English evangelical poet William Cowper had London in mind when putting forward the dictum that has long characterised attitudes towards the city: 'God made the country, and man made the town' (Cowper, 1785, book i).

This indicates the prevailing lines of the debate, despite the fact that 'man' made the country no less than the town (Porter, 1994, 41). Town and country were represented as the opposites of one another. The town was regarded with suspicion because it subjected its inhabitants to an unnatural existence, from which the effects could only be adverse. One can easily see the reasons for the mistrust. In Britain (excluding Ireland) the population trebled in the nineteenth century. In one estimate, 21.3 per cent of the English had lived in towns of 20,000 or more inhabitants in 1801, but the figure had reached 61.7 per cent by 1891 (Coleman, 1973). London grew from 1.1 million in 1801 to 2.36 million by 1851, with smaller industrial towns showing even more spectacular rates of growth. Urbanisation, the relative growth of the population living in towns rather than rural areas, took hold. Until the advent of industrialisation, the vast majority of people lived on the land, on farms, in villages, or in small country-towns. These places were considered to offer the basis for a stable existence rooted in the virtues of family and community life, notwithstanding the appalling picture of rural squalor that emerge from texts such as the Reports of the Poor Law Commission. Rural living was deemed close to nature and hence God, and was more likely to lead to moral and spiritual well-being.

This arcadian view of the countryside (see Chapter 4) contrasted with the towns and cities with their dark and enclosed spaces, overcrowding, tension, noise and dirt. Witnessing this unprecedented and largely unplanned 'urban explosion', observers steadfastly believed that the strains of urban living must impact on the quality of human existence. They also believed, with some justification, that the segmentation of society in the Victorian city was so great that the literate classes had little knowledge of living conditions that most people endured. The result was a mid-nineteenth-century torrent of campaigning books, tracts, essays, collections of engravings and photographs in which novelists and others sought to alert their audiences to the physical, social and moral decay that they found. The following exercise provides an opportunity to carry out a more detailed analysis of a specific example.

Exercise 7.1

Consider the following representation of London in the mid-nineteenth century that appears at the start of Charles Dickens' book *Bleak House* published in 1853.

Make a list of the key themes to emerge from this passage.

Why do you feel Dickens creates such a negative portrait of life in the Victorian city?

LONDON. Michaelmas Term lately over and the Lord Chancellor sitting in Lincoln's Inn Hall. Implacable November weather. As much mud in the streets as if the waters had but newly retired from the face of the earth, and it would not be wonderful to meet a Megalosaurus, forty feet long or so, waddling like an elephantine lizard up Holborn Hill. Smoke lowering down from chimney-pots, making a soft black drizzle, with flakes of soot in it as big as full-grown snow-flakes – gone into mourning, one might imagine, for the death of the sun. Dogs, undistinguishable in mire. Horses, scarcely better; splashed to their very blinkers. Foot-passengers, jostling one another's umbrellas, in a general infection of ill-temper, and losing their foot-hold at street-corners, where tens of thousands of other foot-passengers have been slipping and sliding since the day broke (if the day ever broke), adding new deposits to the crust upon crust of mud sticking at those points tenaciously to the pavement, and accumulating at compound interest.

Fog everywhere. Fog up the river, where it flows among green aits and meadows; fog down the river, where it rolls defiled among the tiers of shipping, and the waterside pollutions of a great (and dirty) city.

(Dickens, 1853)

Your list might include the following:

- The weather: rain and fog or, more accurately, the smog of an industrial city.
- The pollution: the mud-spattered streets, the polluted river ('waterside'), and the soot- and smoke-laden atmosphere.
- The animals, spattered with mud.
- The ill-tempered people, who struggled to cope with the demands of this difficult and seemingly alien environment.
- The references to flooding, containing the suggestion of Noah's Flood, and to the megalosaurus – a type of dinosaur.

No novelist did more than Charles Dickens (1812–72) to convey the character of mid-Victorian London – the setting for many of his books. *Bleak House*, which first appeared in monthly instalments in 1852–3, provided powerful depictions of the living conditions of a city with which he was well acquainted. This scene-setting extract has its context in the general representation of cities in art and literature. For example, Dickens' close contemporary, the novelist and Anglican clergyman Charles Kingsley (1819–75) provided similar portrayals in his novel *Alton Locke* (Kingsley, 1850). As part of the explanation of how the hero comes to be socially aware, the author described a companion taking Locke on a walk through Clare Market to St Giles' to shock him into seeing the conditions of London's poor:

It was a foul, chilly, foggy Saturday night. From the butchers' and green-grocers' shops the gas-lights flared and flickered, wild and ghastly, over haggard groups of slipshod dirty women, bargaining for scraps of stale meat and frostbitten vegetables, wrangling about short weight and bad quality. Fish-stalls and fruit-stalls lined the edge of the greasy pavement, sending up odours as foul as the language of sellers and buyers. Blood and sewer-water crawled from under doors and out of spouts, and reeked down the gutters among offal, animal and vegetable, in every stage of putrefaction. Foul vapours rose from cowsheds and slaughter-houses, and the doorways of undrained alleys, where the inhabitants carried the filth out on their shoes from the back-yard into the court, and from the court up into the main street; while above, hanging like cliffs over the streets – those narrow, brawling torrents of filth, and poverty, and sin – the houses with their teeming load of life were piled up into the dingy, choking night. A ghastly, deafening, sickening sight it was.

(Kingsley, 1850, 83–4)

Dickens' depiction of London in November 1853, therefore, echoed what other writers were saying about the condition of the city's working-class areas. Readers would have needed little persuasion to accept such ideas as part of the story line. Dickens' portrayal of city life in *Bleak House* also had a context in the development of the author's own work. As a young man, Dickens avidly consumed London's theatre life nearly every night of the week, ranging from the major theatres to penny music halls (Schwarzbach, 1979, 23). He enjoyed tramping the city streets particularly at night, often covering 30 miles in a single walk. As his daughter Kate recalled: 'He would walk through the busy, noisy streets, which would act on him like a tonic and enable him to take up with new vigour the flagging interest of his story and breathe new life into its pages' (quoted in Schwarzbach, 1979, 27). His early writings in *Sketches by Boz* (Dickens, 1839) saw the observer 'walking the streets of London as if beating the bounds of his domain, marking a writer's territory he was to hold down for nearly four decades' (Hemstedt, 1996, 215). More recently in *Dombey and Son* (Dickens, 1848), Dickens had provided a portrait of London as centre of a world of endless possibilities for human development alongside that of being a place that was inimical to human life.

Yet by the time Dickens wrote *Bleak House*, he had set such ambiguities aside. In a letter to the novelist Bulwer Lytton in 1851, Dickens observed: 'London is a vile place, I sincerely believe. I have never taken kindly to it, since I lived abroad. Whenever I come back from the country now and see that great heavy canopy lowering over the housetops, I wonder what on earth I do there except on obligation' (quoted in Schwarzbach, 1979, 114). Not surprisingly perhaps, the London of *Bleak House* is undoubtedly the 'vile place':

- The prose quickly built up a picture of a dirty and polluted city, in which even the forces of nature were subsumed by artifice.
- The smoke and soot spewed into the atmosphere by household fires created their version of a snowstorm. The fog itself was 'defiled' by the city's pollution.

- The city dislocated the life of its inhabitants, forced to pick their way through the mud-spattered streets and jostling each other's umbrellas. The environment seemingly exerted a powerful hold over the behaviour of the city's inhabitants, who reacted as detached individuals rather than members of a community.
- The mention of the waters being only 'newly retired from the face of the earth' implied a Biblical metaphor. It alluded to the story of Noah and the Flood, the event that in Judeo-Christian theology marked the start of the modern world.
- Interestingly too, in this pre-Darwinian era, there was reference to the reappearance of an antediluvian animal in the shape of a megalosaurus – a type of dinosaur and a metaphor for the city itself. For all London's vaunted civilising virtues, an older primeval world apparently lurked just below the surface.

Dickens' description of London gained further context from its position within the narrative. As the book progressed, Dickens developed elements in this passage. The fog was a constant metaphor for what was real, what was obscured and the changing relationship between the two. It blurred the specific places mentioned into a seemingly endless mass of architecture and ill-defined spaces (Wolfreys, 1996). The emphasis on the mud found everywhere in the streets was part of a developing sub-narrative that alerted the reader to the city's appalling lack of sanitation. As Schwarzbach (1979, 124) noted:

> The mud of mid-century London . . . was compounded of loose soil to be sure, but also of a great deal more, including soot and ashes and street litter, and the fecal matter of the legion horses on whom all transport in London depended. In addition, many sewers (such as they were) were completely open, and in rainy weather would simply overflow into the streets. Dogs, cattle in transit either to Smithfield or through the town (many dairies were still inside the city), and many people as well used the public streets as a privy, but then even most privies were simply holes in the ground with drainage into ditches or another part of the street. . . . The mud must at times have been nothing less than liquid ordure.

Seen in this light, the descriptions of mud testified to Dickens' determination to alert his readers' attention to sanitation in the city and mirrored the author's growing interest in housing reform. In 1850 Dickens founded and edited the magazine *Household Words*, the stated purpose of which was 'the raising up of those that are down, and the general improvement of our social condition' (Welsh, 1971, 15). At the time of writing *Bleak House*, he and his associates were commissioning a model housing project at Bethnal Green in London's East End. Resonances of these wider interests permeate the book.

Dickens' writings had a major impact on contemporary readers, not least in North America where he toured on several occasions. Some indication of that influence is shown in the work of the New England novelist and short-story writer Nathaniel Hawthorne (1804–64).

Exercise 7.2

Read through the next extract, which is taken from *Passages from the English Notebooks* by the American author Nathaniel Hawthorne. Note down the key themes that he mentions.

Now re-read the extract from *Bleak House* above. What similarities do you find? Suggest why these similarities exist?

I have walked the streets a great deal in the dull November days, and always take a certain pleasure in being in the midst of human life, – as closely encompassed by it as it is possible to be anywhere in this world; and in that way of viewing it there is a dull and sombre enjoyment always to be had in Holborn, Fleet Street, Cheapside, and the other busiest parts of London. It is human life; it is this material world; it is a grim and heavy reality. I have never had the same sense of being surrounded by materialisms and hemmed in with the grossness of this earthly existence anywhere else; these broad, crowded streets are so evidently the veins and arteries of an enormous city. London is evidenced in every one of them, just as a megatherium is in each of its separate bones, even if they be small ones. Thus I never fail of a sort of self-congratulation in finding myself, for instance, passing along Ludgate Hill; but in spite of this, it is really an ungladdened life to wander through these huge, thronged ways, over a pavement foul with mud, ground into it by a million of footsteps; jostling against people who do not seem to be individuals; but all one mass, so homogeneous in the street-walking aspect of them; the roar of vehicles pervading me, – wearisome cabs and omnibuses; everywhere the dingy brick edifices heaving themselves up, and shutting out all but a strip of sullen cloud, that serves London for a sky, – in short, a general impression of grime and sordidness; and at this season always a fog scattered along the vista of streets, sometimes so densely as almost to spiritualise the materialism and make the street resemble the other world of worldly people, gross even in ghostliness.

(Hawthorne, 1870, vol. 2, 368–9)

Your list of common themes might include:

- Time and place: the Ludgate Circus area of London, again in November.
- Environmental conditions: crowded streets, grime, mud ground into the pavement, noise of vehicles, dingy buildings.

- Weather: overcast, fog.
- People's behaviour: jostling, weary, ill-tempered; atomised individuals rather than a community.
- Dinosaur metaphor – on this occasion a megatherium.

Nathaniel Hawthorne served as American consul in Liverpool and Manchester between 1853 and 1857. While in post, he often visited London. He penned this extract in his notebook on 6 December 1857, shortly before his return to the USA and, significantly, some four to five years after Dickens wrote his opening to *Bleak House*. Even a cursory glance shows that Hawthorne, to say the least, was strongly influenced by Dickens' writing. Superficially, the similarity might suggest what would now be termed plagiarism. Before the British Government introduced the Copyright Act in 1911, authors freely appropriated the work of others as their own. Hawthorne's prose followed the same lines as that by Dickens, contained similar ideas and offered almost identical representations of people and place.

Closer scrutiny, however, leads to different conclusions. Hawthorne visited the area at the same time of year because he wanted to see it through Dickens' eyes. *Bleak House* effectively served as Hawthorne's guidebook, just as cultural tourists today visit places because of their literary connections. It directed Hawthorne's gaze when he visited this part of the City of London and set a mood that he tried to capture in his writing. Judgement of this piece of text changes, too, when realising how it came to be published. It was found in Hawthorne's notebooks after his death and edited by his wife Sophia into a volume of posthumous papers (Hawthorne, 1870). Commenting on her editorial decisions, Sophia stated that she had omitted some pages of observations because they were 'too personal with regard to himself or others, and sometimes because they were afterwards absorbed into one or another of the Romances or papers . . .' (Hawthorne, 1870, vol. 1, vi). What appears in the extract above was a fragment from an author's notebooks, perhaps written with professional reasons in mind but not intended for publication. One may speculate that its similarities to Dickens' piece was that it was a writing exercise, completed in the same way that visitors to an art gallery make copies of paintings that they like. In the process, Hawthorne developed some of Dickens' ideas. The reference to the dinosaur, for example, was now to the animal's skeletal organisation rather than an ironic allusion to its physical reappearance. Hawthorne also made clear his enthusiasm for London's street life, to the point of even enjoying some of the things that Dickens decried. Two days later (8 December 1857), the *Notebooks* showed Hawthorne returning to the same physical and intellectual territory, but drawing different conclusions:

> I went home by way of Holborn, and the fog was denser than ever, – very black, indeed more like a distillation of mud than anything else; the ghost of mud,

– the spiritual medium of departed mud, through which the dead citizens of London probably tread in the Hades whither they are translated. So heavy was the gloom, that gas was lighted in all the shop windows; and the little charcoal-furnaces of the women and boys, roasting chestnuts, threw a ruddy, misty glow around them. And yet I liked it. This fog seems an atmosphere proper to huge, grimy London; as proper to London as that light neither of the sun or moon is to the New Jerusalem.

(Hawthorne, 1870 vol. 2, 385–6)

Yet despite these differences, the extracts from Dickens and Hawthorne still testify to the consensus about the Victorian industrial city. Dickens, as we noted above, built on a well-established tradition that maintained that the city subverted fundamental values and distanced people from rurality and spirituality. Hawthorne willingly accepted the same sentiments as part of the taken-for-granted character of urban life. Guided by Dickens, his gaze fell upon the grim and squalid aspects of the city's environment and their supposed alienating properties. These had already become standard attributes in representing the city by the mid-nineteenth century and the legacy would prove pervasive. The dominant discourse in literature, photography, film and the broadcasting media continues to this day to emphasise the dark side of urban life and express the view that cities are problem-ridden, perhaps irredeemably so.

At one level, the prevalence of detrimental, even hostile representations of the city and urban life may seem unimportant – simply a matter of artistic choice or intellectual preference. At other levels though, it assumes considerable importance. The ways that cities are represented express deeply held cultural beliefs and expectations shared by academics, literati, artists, professional planners, architects and political decision-makers alike. For all the veneer of professional detachment, those who look for problems are ideologically predisposed to see and represent problems. It may certainly be argued that this type of thinking colours decisions made at all levels, from choice of subjects to research to ways of managing the urban environment.

The question of neighbourhood

I don't think New York is like other cities. It does not have character like Los Angeles or New Orleans. It is all characters – in fact, it is everything. It can destroy a man, but if his eyes are open it cannot bore him. New York is an ugly city, a dirty city. Its climate is a scandal, its politics are used to frighten children, its traffic is madness, its competition is murderous. But there is one thing about it – once you have lived in New York and it has become your home, no place else is good enough.

(John Steinbeck, quoted in Stern et al., 1995, 13)

It was difficult to escape the powerful critique of the industrial city, but there have always been those that have viewed the great cities of North America and Europe in a different light and enthused over their historic buildings, their arts and cultural institutions, their economic vibrancy or their progressive spirit and vivacity. Most often, however, the attraction lies in *places* within the city rather than the city *per se*, celebrating the local scale and the benefits that locality confers. New York is a good example. The praise given by the novelist John Steinbeck (1902–68) was somewhat backhanded, but gained force for loving the city despite its manifest problems. Many had felt much the same, like the poet Walt Whitman (1819–92). Born in Long Island, Whitman spent his childhood and adolescence in Brooklyn, before moving to Manhattan to pursue a career in journalism in 1841. Whitman profoundly believed in the equality of human beings and in the potential of the USA as a stronghold of democracy; significant beliefs in a country then on the verge of sliding towards civil war. His poem 'Crossing Brooklyn Ferry', originally written in 1856, reflected on the city as seen from the deck of the old Brooklyn Ferry (which connected Brooklyn and Manhattan before the Brooklyn Bridge was completed in 1883). The poem was about New York, although the city became a symbol for more general thinking. It concluded:

> Thrive, cities – bring your freight, bring your shows, ample and sufficient
> rivers,
> Expand, being than which none else is perhaps more spiritual,
> Keep your places, objects than which none else is more lasting.
> You have waited, you always wait, you dumb, beautiful ministers,
> We receive you with free sense at last, and are insatiate henceforward,
> We use you, and do not cast you aside – we plant you permanently within us,
> We fathom you not – we love you – there is perfection in you also,
> You furnish your parts towards eternity,
> Great or small, you furnish your parts toward the soul.
> (Whitman, 1975, 195–6)

These words bring together three important themes: the city's dynamism, its ability to embrace all comers (products, immigrants, ideas) from all parts of the globe in a show of unity, and the sense of the parts that made up the greater whole. New York, the key port for North American immigration, was the 'city of final destination . . . the ultimate city of migrants in a nation of migrants' (White, 1949, 121). It was a natural gathering place for people, products and ideas, but its soul remained linked to the parts from which it was composed. An essential element in appreciating the city's soul, therefore, lies in appreciating the nature and role of those 'parts', especially its patchwork of neighbourhoods.

New York, like other large US cities, has a legacy of ethnic and social neighbourhoods reflecting the city's history of immigration and social segmentation.

Writers and filmmakers, who themselves often live in the city, have celebrated the cultural richness of the city's matrix of ethnic neighbourhoods. Pre-1919 immigration saw the formation of Little Germany, Little Italy, Chinatown, the Irish and Jewish neighbourhoods and the rest. Post-1945 immigration saw large numbers from the Middle East, Latin America, the Caribbean (especially Puerto Rico and Haiti), Southeast Asia, Africa and the Balkans settling in neighbourhoods such as the Lower East Side, Flushing, Bay Ridge, Fordham and Jackson Heights in Queens (Day, 2003). Stephen Low's evocation of immigrant life in New York in the IMAX film *Across the Sea of Time* (1995) had the main character Leopold writing home:

> The neighbourhoods sometimes seem like small countries bordered by the city streets. But even with our different tongues and customs, we can still become friends because we are all Americans. . . . Eating here is like travelling around the world, and you can smell where you are before you get there.
>
> (quoted in Neumann, 2001, 112)

New York became Europe in microcosm for people that had left, or had been driven out of their homelands, but took on a utopian aura (see Chapter 8) by its ability to smooth out the divisions and hostilities that plagued the Old World (Neumann, 2001, 112).

In a less rose-tinted but still positive analysis, Jane Jacobs celebrated the grittier elements of New York's neighbourhoods in *The Death and Life of Great American Cities* (Jacobs, 1962). Although her book addressed national planning issues, it focused on Manhattan and particularly Hudson Street where she lived. Jacobs noted how attachment to neighbourhood enhanced security in New York's West Village district. Hudson Street comprised three- and four-storey houses, with stores on the ground floor when the houses had not been converted to single-family use. Jacobs praised the animated nature of the streets, the civility that people showed and the idea that they looked out for one another. In a city increasingly concerned about violence, she concluded: 'We are the lucky possessors of a city order that makes it relatively simple to keep the peace because there are plenty of eyes on the street' (Jacobs, 1962, 54).

Jacobs' spirited endorsement of the neighbourhood met with support from elsewhere. Thinking about the area of Chicago's South Side where he grew up in the 1960s and 1970s, Carlo Rotella (2003, 87–8) identified:

> a quality of civic life and of inner life, a feeling of relation to people and place, that is sustained or destroyed through the statistically unmeasurable processes of culture. To think of someone – even an enemy – as a neighbour acknowledges an obligation or regard not always extended to strangers; to think of a piece of the city as one's neighbourhood is to acknowledge an investment in it that goes

beyond rents or mortgages. The terms and potential consequences of that intangible investment can be found in the way people act toward one another, the ways they imaginatively inhabit the landscape, the ways they think and talk and write about their neighbourhoods, the stories they tell.

In Boston, Walter Firey (1947) recognised that the residents of the upper middle-class Beacon Hill district shared deep attachments to where they lived; attachments that were seemingly strong enough to have allowed the area to withstand pressures for redevelopment. In the same city, a series of studies manifested residents' genuine sadness when forced to relocate through comprehensive redevelopment of their neighbourhoods. Mark Fried (1963; also Fried and Gleicher, 1961), for example, showed that the working-class community in Boston's West End was proud to belong to the area and valued its warm social relationships. The locality effectively became an extension of the home, an area with which people were personally identified. When their neighbourhood was destroyed by the clearance machine, Fried reported a collective feeling of grief, which he compared to that individuals suffer at the loss of a loved one.

Possible representations of rundown inner city neighbourhoods as valued space and having the potential for warm and intricately organised social networks, however, did not accord with the spirit of the times. The 1960s revealed American society's growing doubts about the future of the cities. Those doubts were crystallised and encapsulated by the Kerner Report (1968) on the inner city riots of the 'Long Hot Summers' of the mid-1960s. Set up in July 1967 while rioting was still underway in Detroit, the 11-member National Advisory Commission on Civil Disorders was charged with explaining the riots in the black sections of many major US cities since 1965. Its key finding warned that the USA was 'moving toward two societies, one black, one white – separate and unequal'. The cities emerged as ones in which the white middle classes had fled to the suburbs, leaving behind a declining population with an increasing proportion of black people. The problems that fuelled the riots were recognised as persistent unemployment, poor housing, poor education opportunities, lack of access to health care, and unremitting racial discrimination – all of which particularly affected black Americans. The report also revealed evidence of systematic police bias and brutality against black communities.

The powerful language used in the Kerner Report, and the images that these terms carried, were reinforced by a new wave of studies that questioned the role of neighbourhood in American city life. Studs Terkel's oral historical approach in *Division Street, America* (Terkel, 1966) used findings from Chicago that had already suggested ethnic neighbourhoods were breaking down as those with resources fled to the suburbs:

> Chicago's large ethnic groups – Poles, Czechs, Yugoslavs, Ukrainians, Germans, Scandinavians, Irish, Italians, Jews – no longer live in such homogeneous areas as formerly. There are still a few such areas to the south and west – and Germans and Scandinavians to the north – but the flavour is no longer lasting. Most of the younger generation of these groups have moved to the suburbs.
>
> (Terkel, 1966, 22)

By contrast, those excluded from that process by socio-economic forces and racial discrimination, remained *in situ*:

> Much of the South Side and Near West Side comprise the black ghettoes, the most expansive in the country. There are smaller such ghettoes in other areas of the city. This is unofficial, of course. Mayor Richard J. Daley, on 4 July 1964, proclaimed: 'There are no ghettoes in Chicago'.
>
> (Terkel, 1966, 21)

To Terkel's respondents, neighbourhood – even for the South Side neighbourhood to which Rotella referred above – was a residual category, the existence of which signified exclusion and inequality of opportunity. Others saw its value primarily in terms of raw territoriality. Robert Gold (1971), for instance, argued that American cities had assumed a defensive and cellular structure, with separate and unequal groups clustering together in urban space, displaying hostility towards potential intruders. Gerald Suttles (1972, 21) expanded this theme by identifying 'defended neighbourhoods' as the 'residential group which seals itself off through the efforts of delinquent gangs, by restrictive covenant, by sharp boundaries, or by a forbidding reputation'. In this way, he identified the defended neighbourhood as extending beyond the immediate context of the inner city to embrace areas of exclusive housing. The title of David Ley's monograph *The Black Inner City as Frontier Outpost*, which concerned an inner-city neighbourhood in Philadelphia, caught the mood of the moment (Ley, 1974). In the process, the value of neighbourhood was seen primarily as offering a safe haven in the otherwise hostile environment of the city.

In due course, anxiety about the city lost its apocalyptic edge, but the idea of neighbourhood as a safe haven still resonated in housing markets. Stress on 24-hour security and freedom to enjoy the good life behind protected boundaries became selling points for property developers and real estate agents. In their uncompromisingly titled *Fortress America*, for instance, Blakely and Snyder (1997) examined the growing tendency for communities in the USA to cluster together behind walls and boundaries. These 'gated communities' are particularly associated with suburban developments, but also breed a new texture back into the city through establishing 'security zone communities'. Fear rather than warm neighbourliness underpins the way that these urban environments form:

The drive to redefine territory and protect boundaries is being felt in neigh-bourhoods of all income levels throughout America's cities. Much of the growth in gated communities is not created by developers but by residents of existing neighbourhoods who install gates and barricades in an attempt to defend their existing way of life. These are the security zone communities, the closed streets of the city, suburb, and barricade perches. We define this type by the origin of its gates and fences: unlike the lifestyle and prestige communities, where gates are built by the developer, in security zone communities the residents build gates and retrofit their neighbourhoods with security mechanisms. In the city and suburb perches, residents turn their neighbourhoods into gated communities by closing off all access and sometimes hiring guards.

(Blakely and Snyder, 1997, 99)

Following these arguments, what is portrayed as good city life is often a critical view of the city *per se*. Again, the defended neighbourhood is an oasis or a retreat; the city beyond remains a lifeless or hostile place (see also Davis, 1992).

Not all developments, however, point in this direction. The Safer Cities pro-grammes in the USA, for example, attempted to build on and extend notions of neighbourhood security in the 1990s. In Chicago, the Alliance for Neighbourhood Safety, a coalition of more than 30 neighbourhood organisations, developed a community-policing proposal for the city. Boston's Safe Neighbourhoods Programme saw areas working together to target gang violence. St Paul (Minnesota) introduced neighbourhood initiatives aimed at domestic violence, reduced property and car crime (Wekerle and Whitzman, 1995, 11). Newark (New Jersey) involved neighbourhood co-operation in targeting gun crime. These and similar schemes (Brunn *et al.*, 2000) build on a sense of neighbourhoods as distinctive social areas that again can work together to improve law and order in cities.

Exercise 7.3

In this section on 'the question of neighbourhood', we have seen the urban neighbourhood represented in a variety of ways. Look back through this material and make a list of the different ways that 'neighbourhood' is represented.

To what extent can each of these representations be classified as sympathetic or hostile?

Your list should include the following:

• An area with a distinct and valued identity, often conferred by its immi-grant past, possessing warm social relationships and acting as a tool for democracy.

- A residual area, left behind when those with greater resources fled to the suburbs and suffering exclusion, deprivation and high crime rates.
- A secure haven, a building block for community development in an insecure urban environment – an idea much favoured by property developers and real estate agents.
- A tool for regenerating co-operation and law-abiding behaviour.

With regard to the underlying values expressed, it is possible to think of these as comprising a spectrum ranging from hostility to romantic sympathy. At one end came a view of neighbourhoods as relics, the places occupied by those lacking the means to flee to the suburbs. Between these extremes came portrayals of neighbourhoods as places for breeding security and greater co-operation back into the lawless city. At the other end of the spectrum came the romantic conception of neighbourhoods as places with warm community relationships that acted as crucibles of identity and democracy.

These modes of representation are sometimes applied simultaneously to the same area. To some the run-down inner-city neighbourhood may well be a menacing and deprived place; to others it may have compensating benefits; and to others it may be a valuable investment opportunity once refurbishment and the apparatus for physical security are installed. Moreover, the list of possible representations continues to expand. Current debates over urban sustainability (see Chapter 8) have highlighted the value of the existing city and possible ways of reanimating its neighbourhoods as social units. Representation, as before, is a dynamic process.

Place promotion

If enthusiasm for the city as the setting for the Good Life is generally hard to find, there remains one field in which the discourse of pro-urbanism flourishes as never before; namely, the multi-million pound industry of city marketing and, specifically, the realm of place promotion. *Place promotion*, defined as the conscious use of publicity and marketing to communicate selective images of specific geographical localities or areas to a target audience, has a long history (Gold and Ward, 1994). It began in colonial days as a part of the strategy for peopling the wilderness, in which context it served two distinct and related functions. First, promoters needed to tempt sponsors to invest in risky schemes on the very margins of Empire or to induce migrants to take their families into the unknown. To do so, they resorted to a predictable litany of advantages to show that such colonies offered unparalleled opportunities and unequalled bounty for the intrepid. Second, and partly as a result of the consistent failures of colonies to live up to the extravagant claims made for them, meant that promoters needed to counter the 'enormities and abuses' heaped on such developments in case these seriously dented the potential viability of the colony. If that meant embroidering the truth, inventing favourable dispatches, or selectively presenting the reports of explorers or surveyors in advertisements, then so be it (Jones, 1946; Cameron, 1974).

This two-edged strategy of making claims and countering stereotypes remains a standard feature of place promotion to this day, especially in support of town development. At first, particularly in the USA, this involved simple strategies such as direct advertising for migrants or, less directly, the nicknaming of towns and cities to give them a distinctive identity. For instance, before the fire of 1871, Chicago represented itself as the Garden City and the Gem of the Prairies. Later, as its self-conception altered, the city boosters adopted other names. While others might have called it the Windy City, City of Big Shoulders and the Crime Capital, Chicago officials stressed its new economic role by calling it the Metropolis of the West, Hogopolis, Cornopolis, the Hub of American Merchandising and the Country's Greatest Rail Centre (Tuan, 1974, 202–3). Over time this activity took on a sharper edge as place promotion was drawn into broader strategies of urban and regional development. In his survey, Ward (1998) examined the comparative progress of locally based place promotional activity in Britain, the USA and Canada from 1870 onwards. He noted that differing economic, institutional and political circumstances have shaped promotional policies in various ways. Thus, while campaigns to attract investors and visitors became a regular and accepted part of local economic policy in the northeastern states of the USA in the 1840s, the same policy did not spread to Canada until the later decades of the nineteenth century, and to Britain in the 1930s. In Britain, for example, place promotion was primarily oriented towards tourism, particularly seaside resorts, and residential developments – notably the new suburbs.

Selling the interwar suburb

Traditionally suburbs were considered *sub*-urban. In medieval towns, for example, they comprised the poorest, most squalid and ramshackle dwellings (Thompson, 1982, 2). By the early twentieth century though, suburbia had experienced a profound cultural reappraisal. Changes in the housing market, increases in disposable incomes and the availability of loan capital to purchase houses at low rates of interest saw the professional middle classes clamouring for houses in the suburbs. This was particularly true in the London region. With its role as the administrative capital of nation and Empire, its nodal position in an expanding communications network and its rapidly expanding service sector as well as its manufacturing base, London grew from 6.5 million in 1914 to 8.5 million by 1939. This increase was accommodated by setting free London's centrifugal tendencies, with the areal extent of the city doubling between 1919 and 1939. The new suburbs engulfed large areas of the adjacent counties, forming a continuous built-up area extending up to 15 miles from the city centre.

This caused considerable alarm among urban commentators and practitioners concerned about the physical spread of suburbia and intellectuals anxious about its potential social consequences. The language used was symptomatic. The suburb was likened to the spread of the 'brown rat, ravaging everything before it' (Robertson, 1925, 68). The semi-detached British suburban house was described as 'perhaps the least satisfactory building unit in the world' (Abercrombie, 1939, xix) with its design being castigated as the 'neo-Nothing' (Nairn, 1961a, 384), bought for the 'social cachet that . . . (it is) presumed to bring' (Nairn, 1961b, 164) and its 'repetition' being held to produce 'an inescapable monotony of mass' (Abercrombie, 1939, xix). Suburban estates, then, were judged failures 'from a sociological as well as from an aesthetic point of view' (Sharp, 1936, 115). The reasons for this outpouring of condemnation lay as much in English class prejudice as in genuine alarm about loss of agricultural land or implications for service provision (Carey, 1992), but it meant that those selling suburban housing needed the familiar two-pronged thrust of place promotion. In other words, as well as devising a selling message that appealed to the new middle-class homebuyers, they also needed to combat the powerful representations of suburbia as a placeless landscape of replication and pretension. The activities of the building societies provide considerable insight about how the task was tackled.

The building societies rose from obscure beginnings as friendly societies and 'building clubs' at the time of the Napoleonic Wars to larger formal groupings later in the nineteenth century. Under section 13 of the 1874 Buildings Societies Act, they incorporated 'for the purpose of raising, by the subscriptions of members, a stock or fund for making advances to members out of the funds of the society upon freehold, copyhold, or leasehold estate by way of mortgage'. For reasons discussed elsewhere (Gold and Gold, 1994), they found themselves awash with funds during the years between 1919 and 1939. Needing to increase the flow of mortgage capital to borrowers, they expanded their networks of branch offices and started promotional campaigns to persuade more people to become owner-occupiers. At first their efforts were regarded as amateurish and unproductive, but they quickly developed more effective and sustained campaigns. Typical media included local handbooks, extensive newspaper advertising and posters. The posters, as exemplified by Figure 7.1, measured approximately 20 × 30 inches and were designed for display in the windows of building society branches and estate agents. Many were brightly coloured and based on lithographed paintings, photographs

Let's build it
through the

HALIFAX
BUILDING SOCIETY

*Ask for the latest
Homebuying booklet*

APPLY WITHIN

Figure 7.1 *Poster, Halifax Building Society. Courtesy of Halifax plc*

and graphic designs; all intended to be eye-catching at a time when high-street office and commercial façades were overwhelmingly drab.

Exercise 7.4

Figure 7.1 shows a Halifax Building Society poster from the mid-1930s. Study it carefully and write a short description of what the advertisement contains.

Pay particular attention to any representations of *gender* that you find.

Figure 7.1 shows the following:

1   A couple studying a site plan of a new suburban estate on their wooden dining room table. The man, standing, was dressed in a typical three-piece suit, with a handkerchief in the top pocket, suggesting that he worked in an office in the city. The woman, seated, was wearing smart everyday clothes. Given the patterns of family life in the 1930s and given that her wedding ring was just visible, she was probably a housewife.

2   The advertisement expresses typical 1930s views about family life and gender. The husband stands authoritatively while his wife was seated. She held a Halifax Building Society booklet on choosing a home in her right hand, but this was given no attention. The focus of the picture directed attention to the property on the plan to which the husband pointed with the index finger on his left hand. *That* was the one that would do. His right arm was lovingly but protectively placed round the back of her chair. She looked on smilingly, seemingly approving his decision.

3   Behind them, lightly rendered as dream-like background, was the subject of their aspirations: a new house in the suburbs. The house was typical of the period, with tiled roof, steep-pitched gables, casement windows and tile hangings. A radio aerial attached to the chimney stack gave a hint that the suburban house was up-to-date on the inside even if traditional on the outside. The sunlit garden was filled with mature trees and bushes.

4   Due to a device then commonly used, we cannot tell the exact size of the property. This was probably a semi-detached house with a modest garden but the artist used the foliage of trees and bushes to conceal its actual size and extent. There was also only a hint of neighbouring houses, even though these were urban developments of significant density. These representations were part of the place promoter's art. The great English country house, with its spacious grounds, represented, and for many still represents, the acme of desirable housing. Such houses were beyond the wildest dreams of the vast majority of the population, but homebuyers could fantasise about the nearest practical alternative, the detached suburban house. There was therefore the paradox that the semi-detached house, the overwhelming staple of suburbia, was almost never shown in the building societies' promotional posters. Instead what were depicted were either pictures of detached houses or villas, with their connotations of greater social exclusiveness, or carefully crafted portrayals that disguised semi-detached houses to look like detached. This poster followed the latter strategy. The foliage from the trees and bushes obscured boundary fences or the presence of the attached property. All was done to convey an impression of privacy and spaciousness.

The richness of meaning from this single poster belies the apparent simplicity of its content. The building societies became highly proficient at distilling a package of essential selling messages down to their essence. As such, six distinct, but overlapping, themes stood out as part of that package (Gold and Gold, 1990).

1   *Proximity to nature.* If the town divorced people from their natural sur-roundings, the suburb could be construed as offering a compromise, blending green surroundings and rural tranquillity with the necessary contacts with the

city. In this sense, advertising served a dual purpose. As well as making an argument about reconnecting city dwellers with nature, it also served to counter the gap between reality and potential. At first, there would have been little to see in London's new suburbs apart from unfinished roads rutted by builders' carts and brash new houses rising from a sea of mud. Promotional work therefore invited prospective buyers to look forwards to when the houses were finished, the gardens were flowering, the landscaping complete, the trees mature and the community had developed.

2   *A healthy retreat.* Working in the city might be an economic necessity, but living there was regarded as hectic, stressful and unpleasant. Suburban promoters made much of retreating to an environment that was greener, pleasanter and, by implication, healthier. This idea resonated with 1930s thinking about preventive medicine. Availability of sunlight, fresh air and exercise were seen as essential for individual health, especially for children. Any features of local topography thought likely to promote health featured prominently in promotional work. It was therefore common for much to be made of local topography, aspect, air circulation, microclimate and even soil (favouring gravel soils with their ready drainage of water rather than heavier and sometimes waterlogged London clays).

3   *A place to grow.* Suburban promotion was rich with organic metaphors. The healthiness of the environment was coupled with the greater security of suburban estates, where children could wander freely away from traffic and possible malign influences. Advertisements frequently used the gendered representation of a small girl playing unattended in the suburban garden to convey this sense of greater safety.

4   *Environment and social status.* We noted above that graphic designers almost invariably used pictures or photographs of atypical detached villas to give suburban developments a greater social cachet. For their part, copywriters also devised techniques to suggest that this was a socially exclusive environment. These included:

- emphasising the detachment of the estate from other less desirable areas;
- stressing that the houses were built on what had previously been a landed estate;
- showing that individuals with social standing – especially professionals such as solicitors (attorneys), doctors or architects – chose to live there.

In effect, every effort was made to persuade the likely house buyer that a move to the suburb represented a move up the social ladder as well as access to better housing.

5   *A suburb amongst suburbs.* Despite these entreaties, promoters knew well enough that there was prejudice against suburbia and suburban life. Just as

promoters of new settlements realised that they had to counter stereotypes (see Chapters 2 and 5), so the house builders and building societies understood that they needed to counter ideas that could jeopardise their chances of selling properties. One method was to work with the grain of the argument, implicitly accepting prejudices against suburbia, but maintaining that this suburb was different. House builders tried to give houses greater identity by addition of design subtleties in external finish, such as rendering, beaming, coloured bricks, tile-hangings, plaster work and other ornamentation. The building societies stressed these differences and any distinctive features of layout to help convince people that their new estate was set apart from the norm of suburbia. Indeed, it was noticeable that more prestigious synonyms like 'park estate' or 'garden village' were preferred to the word 'suburb'.

6   *Modernity within limits.* Suburban promoters pitched their product firmly within the traditions of gracious English housing, yet they were also selling modernity. Whatever the traditional references in exterior design, internally these represented a break with the past, with fitted kitchens, bathrooms, separate bedrooms for children and even central heating. Although it was rare to find the aesthetics of modern architecture in advertising, much was made of 'modern convenience' and the quality of fixtures and fittings. Tradition need not mean lack of proper amenities; modernity could be celebrated within limits.

Using these basic themes, the building societies quickly built up campaigns that articulated and endlessly reinforced a carefully chosen repertoire of environmental representations to communicate both the idea of home-ownership and the desirability of the new suburbs to their audience. The evidence, as measured by the unprecedented growth of suburbia during the interwar period, suggests that the building societies had accurately gauged the housing dreams and aspirations of the middle classes. Their artists and graphic designers clearly identified a set of values relating to family and property that English society held most dear (Fishman, 1987) and incorporated them into their selling message. In doing so, they produced a language of images and symbols that were neither an exact reflection of reality nor a fairground distorting mirror, but a mirror that distorted, selected and enhanced. The historic success of that message strongly suggests that the audience fully understood that language.

Promoting the cities

In the period since 1945, economic and industrial place promotion progressively became a ubiquitous part of urban and regional policy – even in Britain. It was initially connected to the promotion of growth in depressed regions and planned expansion elsewhere. It was an era in which almost anywhere claimed to be the centre of ____, heart of ____, or gateway to ____ (Gold, 1974; Short and Kim, 1996). These simple emphases on place identity and centrality were also basic ingredients of advertising. Promoters endlessly recycled packages of ideas about their towns that included 'superior' location, good labour relations, ample supplies of building land, excellent transport infrastructure and good access to open country. The logic behind such advertising derived directly from the traditional industrial identity of these towns and regions. They were centres for manufacturing and primarily provided lists of attractions to tempt manufacturing firms to set up factories in their areas.

Change initially came from two sources: the economic problems of the recession-hit 1970s and 1980s, which raised the priority attached to place promotion; and the rise of 'New Right' national governments that encouraged a more entre-preneurial approach to the local management of urban affairs. In other words, against a political climate that stressed individualism and enterprise, cities and regions were seen as bundles of social and economic opportunities competing against one another in the open market for a share of capital investment (Philo and Kearns, 1993, 18). Essential ingredients in that approach were risk-taking by public–private partnerships, profit motivation, inventiveness, place promotion and marketing.

Detailed discussion of the new urban entrepreneurialism, and the switch from urban government to so-called 'governance', lie outside the scope of this study (see Hall and Hubbard, 1996; Harvey, 2001), but the activities of image building, rebranding and place marketing deserve further comment. Ward (1998, 191–2) argued that the key developments occurred in the late 1970s in Boston and New York. Boston's Mayor, Kevin White, began a campaign to create a new image for the city based on 'advertising its commercial and cultural advantages, and drawing the attention of visitors and tourists to (its) virtues as a world class city' (O'Connor, 1993; quoted in Ward, 1998, 191). In the early 1970s, the New York Convention and Visitors Bureau started a promotional campaign by getting comedians and sports stars to hand out little red 'Big Apple' lapel pins (Posner, 2000, 246). Rather more decisively, in 1977 the city launched the 'I ♥ New York' campaign using advertising to create a positive feeling about a place then on the brink of fiscal collapse. Its success was only partly measured by the countless times that its prime slogan was copied elsewhere. Rather it showed that advertising *could*, in the right circumstances, help create positive images of places perceived as 'tired, seedy and declining' (Ward, 1998, 191).

The impact on other cities and their advertising budgets was immediate – even if the funds involved remained modest by the standards of commercial product advertising (Holcomb, 1994). Recession and unemployment-hit cities such as Baltimore, Cleveland, Newark, Pittsburgh and Detroit (see below, p. 206) struggled to create a profitable growth machine focused on tourism, leisure and conspicuous consumption as an antidote to ailing industry, falling profits and urban decline (Harvey, 2001, 143). Each accepted the case for promotional activity, albeit with varying enthusiasm, and poured funds into campaigns aimed at one another, at cities in the Sun Belt and at rivals overseas.

Glasgow, Scotland's largest city and former industrial powerhouse, was an example of a city in the United Kingdom that closely followed New York's lead. It had long suffered from an image of social violence, bad housing and industrial strife epitomised by the novel *No Mean City* (McArthur and Kingsley Long, 1935). Taking a leaf out of New York's book, the city began a reimaging process in 1983 with the launch of the 'Glasgow's Miles Better' campaign. The campaign accented the positive about Glasgow, complete with Mr Happy, a smiling children's cartoon book character, as logo. The campaign lasted from 1983 to 1989 at a cost of £1 million and sought to lay the ghost of the city's industrial past. It was replaced in 1990 by a new programme co-ordinated by the international advertising agency, Saatchi and Saatchi. Timed to coincide with the city's status as European City of Culture, the campaign featured the slogan 'There's a Lot Glasgowing On in 1990' and a 'flying G' logo. Additional publicity media included events brochures, advertising on buses and taxis, umbrellas, badges, T-shirts and widely distributed posters (Gold and Gold, 1995). This, in turn, was replaced by the 'Glasgow's Alive' campaign, which continued to hammer away at the past – that this was no longer a rundown industrial city, that there was vibrant culture, that 'Glasgow's on the move'. In doing so, the city made use of an imagery that is replicated endlessly in place promotion. Cities throughout the world now sell themselves by reference to their cultural attractions, mirror-glass post-modern buildings, newly created central plazas, tastefully converted shopping arcades in their central business districts, and the new marinas in their former docklands (Gold, 1994). Glasgow followed suit. One popular advertisement in its 1992–4 'Glasgow's Alive' campaign was a collage of images aimed at the business market uncompromisingly put forward under the explicit slogan 'Glasgow: No Mean City' (see Gold and Gold, 1995). The wheel had come full circle.

Other cities followed similar routes. Woollongong (New South Wales, Australia) was traditionally an industrial centre focused on steelworks, port industry and manufacturing, and brown coal working. Deindustrialization and downsizing of manufacturing led to sharp job losses, with unemployment at 17.3 per cent by 1994. At this point, the city council decided to try to reinvent the city through development of new industry, education and particularly culture. After consultation, the city launched a six-point plan to encourage cultural activity in 1998 and a five-year campaign in 1999 to change the city's image from 'Steel City' to 'City of Innovation'. Besides urban regeneration schemes involving creation of cultural quarters, the council invented instant tradition in the shape of the 'Viva La Gong' festival. Among other things, this would provide benefits that include: 'promoting . . . local cultural diversity, stimulating a sense of community identity and pride, facilitating partnerships between communities and organisations, (and) focusing . . . creative energy (Buckland, 2002).

Culture and rebranding

The emphasis on culture in these examples highlights an important trend. City authorities throughout the world have recognised that traditional place promotion, with its emphasis on attracting large-scale manufacturing industry, is now largely redundant. While attraction of new high-technology industries still has its place, 'culture' now plays an increasingly vital role in promoting places and in the task of giving cities new images (Gold and Gold, 2005). Treating culture as a saleable product draws on the full range of the term's meanings discussed in Chapter 1. These include:

- *Cultural infrastructure*. The new concert halls, auditoria, state-of-the-art stadia and museums housed in architecturally innovative structures supply signature images that readily transfer themselves to glossy advertisements and slick web pages.
- *Cultural diversity*. Cities now represent themselves as multi-ethnic centres as a way of increasing their tourist flows. Marketing agencies and tourist boards collaborate with previously neglected minority groups to package 'colourful' cultural festivals with associated attractions to draw in tourists.
- Newly defined *cultural quarters*, complete with appropriate street furniture and signs, in a new, market-oriented version of ethnic neighbourhoods. 'Selling points' include regional cuisine, crafts, spiritual centres and 'heritage' (see Chapter 8).
- Incorporation of *cultural industries* (e.g. advertising, architecture, crafts, media and fashion) into new creative clusters.
- *Culture leadership*, signifying sophistication, creative excellence and the respect that a city has on the international stage.

The general points are central to the rebranding exercises that many cities have undertaken. Again launched by New York's 1977 campaign, 'rebranding' treats cities as commodities and seeks to develop an identity based around key ideas, logos and symbols that parsimoniously communicate positive ideas about the product to the outside world. The cities of Hull and Birmingham in England, for example, have recently employed consultants to select visual symbols, logos, common typefaces and colour palettes for use in all communications about the city including letterheads, literature, signage and banners (Guthrie, 2003). Nevertheless, as the head of the consultancy firm that advised Birmingham noted: 'All we have at the moment is a logo and some common tools. That is not the same as rebranding a whole city'. This is perfectly understandable because rebranding poses inevitable problems for fading industrial cities looking for a new identity. The search for a 'saleable product' ignores the fact that cities are complex entities that have always generated diverse and conflicting representations. They contain spatially distinctive areas, offer different living and working environments, and house constituent groups that wish their interests to be recognised and included in the promotional mix. Birmingham's minority ethnic groupings, for example, have long felt ignored by the city (Dudrah, 2002). The city's new slogan 'Many worlds . . . one great city' apparently referred to that diversity, although the slogan 'is queasily reminiscent of the hack travel writer's description of anywhere from Ecuador to Luxembourg as a "land of contrast"' (Guthrie, 2003). Quite simply, cities are diverse phenomena, whereas rebranding often requires a single-minded, or possibly simple-minded clarity and a ruthless pruning of representational diversity that it may be neither practical nor politically possible to achieve. Detroit (Michigan) provides a good example of these points.

Detroit

Detroit's roller coaster history is spectacular even by the standards of American cities. Founded in 1701 as a French outpost to block British imperial expansion, Detroit passed to the British in 1763 and to the Americans in 1796. During the nineteenth century, it became the financial centre of Michigan's natural resource wealth. In the early part of the twentieth century, the city grew rapidly with the establishment of car factories by Ransom E. Olds, John Dodge and Henry Ford. The population more than doubled from 465,000 in 1910 to more than 990,000 in 1920, making it the fourth largest US city (Daley, 2003, 20). Detroit's place promoters worked hard to capitalise on the change. The 'Fair City of the Straits' (Leake, 1912) quickly developed into the 'Motor City' and the 'Automotive Capital of the World' – indeed the name 'Detroit' was synonymous with the American car industry. Local agencies commissioned guides to the leading citizens of Detroit (e.g. Marquis, 1908) to convince the world of its standing as a superior place to live. Detroit's wartime success in switching to munitions production earned it the nickname of the 'Arsenal of Democracy'. An article in *Time* magazine in 1951, celebrating the 250th anniversary of Detroit's foundation, claimed that it best represented the spirit of twentieth-century America, exemplified in its mass production, drive, energy, purpose, and fusion of people and machines (Neill, 1995, 118). The city's continuing success seemed assured.

The assurance, however, masked long-standing social and racial problems as well as economic decline. The *City of Destiny* (Stark, 1943) quickly became the *City Primeval* (Leonard, 1980) and *Pariah City* (Neill *et al.*, 1995). Riots in June 1943 marked the ugly climax to increasingly bitter racial conflicts over jobs and housing. Eventually quelled by federal troops, the riots left 34 dead and 675 injured, primarily Black Americans (Fine, 1989, 1). As shown in Table 7.1, after reaching its high point of 1,848,568 in 1950, the population fell during each subsequent intercensal period to stand at 951,270 in 2000. During the same period, the percentage of Black Americans in population rose from 16 to 75 per cent.

Table 7.1 *Population statistics for central Detroit, 1950 and 2000*

	1950	2000
Population	1,848,568	951,270
% Black	16	75

Source: Based on Fine (1989, 3) and Neill (1995, 113)

These changes accentuated the city's racial problems as the white population fled to the suburbs – Detroit losing almost a quarter of its white population in the 1950s alone. Economic decline hit the downtown area, leaving many of the buildings vacant, dilapidated or obsolete (Thomas, 1997, 15). Deindustrialisation left the minority communities increasingly isolated in the urban core. Soaring crime rates earned Detroit the new title of 'Murder City' (Daley, 2003, 20). Irvine Barat (2001, 55) evocatively portrayed life in the inner-city neighbourhoods:

> It's a neighbourhood
> you're glad you don't live in
> tired bars
> hang-around types
> with chronically thin wallets
> girl-women who parade their emptiness
> on a faded street
> peddling the only product they own
> and those who crowd the files
> of caseworkers.

Frustration with this situation and clashes with the predominantly white police sparked off the inner city riots during the 'Long Hot Summer' of 1967 (see also

Kerner Report above, p. 193). With 43 dead, thousands injured and around $200 million in property damage, this event certainly brought the city global prominence but not of a type that it wanted.

The city authorities tried repeatedly to tackle the situation through regeneration schemes backed by promotional campaigns (Farley *et al.*, 2000). These included the opening of the Detroit Renaissance Centre on the riverfront in 1977; a mixed land use, but fortress-like development intended to create a new symbol for the city and feature in its publicity materials. Despite the initial impact of this and other flagship developments, the 1980s saw sufficient deepening of the city's prevalent economic, social and infrastructural problems for the *Washington Post* to describe Detroit in February 1997 as 'the nation's pre-eminent basket case' (Eisinger, 2003, 86). Certainly, the futuristic depictions of the central area as urban wasteland in the *Robocop* science-fiction films remained closer to the public perceptions of Detroit than the efforts of the city's place promoters (see also Chapter 7).

The election of Dennis Archer as Mayor in late 1993, which coincided with an economic boom, saw more determined efforts to 're-envision land use' and generally sell the town (Thomas, 1997, 201). Facing a budget deficit of $80 million and chronically deteriorated city services, Archer's regime began talking up Detroit as an 'entrepreneur's dream' and possible future 'urban jewel' (*The Economist*, 12 March 1994; quoted in Neill, 1995, 156). The associated rhetoric stressed the importance of the appropriate vision to guide planning, but visions include and exclude. In the current political climate, city authorities have less scope for rebranding exercises founded on consultant-approved visionary conceptions of the city in which large sections of the community, and their interests, are invisible.

The most recent manifestations of this point came from the city's latest rebranding campaign, which was intended to stress three main visions of the new Detroit (Eisinger, 2003, 91), namely:

1   Detroit as an emerging 'world class city': Mayor Archer's original gambit claiming that the city is 'on the verge of greatness' rather than harking back to its undoubted but tarnished world status of the past.
2   Detroit as a tourist destination: 2002 slogan, 'It's a great time in Detroit'.
3   Detroit as the vibrant, modern social and business centre of a booming metropolitan region.

Measured by international standards, these 'visions' are completely unexceptional. They are as much part of modern day city marketing as were the emphases on centrality and place identity for an earlier generation's manufacturing-oriented promotional advertising. Unfortunately for Detroit, and many similar industrial cities, these apparently anodyne visions of 'Any City, Anywhere', encounter two significant problems when applied specifically.

First, they have implications for the city taking shape. Promoting Detroit as a tourist centre focuses representation around the regenerated theatre districts and hermetic casinos and stadia, all in a well policed 'tourist bubble' (Judd, 1999) that excludes large sections of the local population. Re-imagining the city as a thriving regional business and shopping centre needs selectivity in the representation of a city that 'is distinguished from other large (US) cities by its exceedingly weak market for downtown office space and retail shopping' (Eisinger, 2003, 87). Propagating Detroit as an emerging world centre requires scouring the city for elements to support what, by any current assessment, is a windy rhetoric. Each is a partial message framed around a specific promotional need, taking fragments of the city and presenting them as the city itself. The risk of perpetuating a dual city is ever present (Eisinger, 2003, 97).

Second, these visions also sidestep the problem of confronting Detroit's past. Detroit was once a 'world class city' and a thriving regional commercial centre, but these reputations were built around economic activities and structures that nowadays scarcely exist. Leaving aside new attractions, a significant part of tourist promotion invited visitors to see a core of historic buildings or neighbourhoods, but they are unreliable indices of a contested past. There would be a case, for example, in providing facilities that dealt with the city's engineering industrial heritage (see also Chapter 7). Any such facility, however, would perforce confront the differing experiences of different sections of the community. As Eisinger (2003, 89) commented: 'If whites remember a city of sturdy neighbourhoods, ethnic festivals and steady work in the auto plants, black Detroiters remember a different place . . . a city of bitter housing segregation, white neighbourhood associations, and racist mayors.'

In the final analysis, questions concerning the representation of past and future Detroit involve complex ideological matters. Should promotional representations merely be shaped by cost-effectiveness and specific redevelopment initiatives in the belief that all will benefit in the long term, or do they need to embody issues of social justice from the outset? Looking at the significance of these issues, John Bukowczyk (1989, 23–4) commented:

> Is [Detroit] a node in the complex American economy, a place to generate corporate profit and, only secondarily and serendipitously, to enhance the welfare of those who reside in its midst? Or is it, first and foremost, the place where large numbers of people live and its principal purpose providing them with justice, security, education, homes, and jobs? Ambiguity on this point is the real 'urban crisis' we are in – in Detroit and other American cities.

Conclusion

The changing representations of Detroit found in this case study sum up many themes of this chapter. It shows how what was once a highly successful industrial city is now struggling to reconstruct and rebrand itself. Each phase described – the initial boosterism of the place promoters, the appearance of deep-seated social and economic problems that severely damaged the city's image and reputation, and the attempts to re-imagine the city – were not just characteristics of Detroit. Rather they resonate with the wider arguments that we have analysed in this chapter about the significance of values in representing the urban environment.

To recap, we began by exploring the reasons for the duality found in representations of the city since ancient times. After noting the parallel existence of pro- and anti-urban discourses until the eighteenth century, we saw how the pendulum swung decisively towards the latter with the onset of industrialisation. Many Victorian authors believed that the state of the city brought misery and degradation to the urban population and left a powerful legacy of writing about its deficiencies as a living and working environment. This, in most respects, is the orthodox version when considering representations of urban environments.

We then examined representations of urban neighbourhoods in American cities; places that generate contrasting feelings among commentators on the city. At one end of the spectrum came the romantic view of the traditional patchwork of ex-immigrant ethnic neighbourhoods, celebrated as hearths of identity and crucibles of democracy. At the other end of the spectrum came a view of neighbourhoods as a residual category; places of deprivation and exclusion left behind by the flight to the suburbs. Between came portrayals of neighbourhoods as places for breeding security and co-operation back into the city.

Although hostility towards the city dominates urban representations, we examined the other side of the coin by looking at the work of city promoters yet, even here, the counter-theme was ever-present. Place promoters often see the need to counter a town's poor reputation as an indispensable part of the process of conveying an alternative, positive message. We saw this in the case study of the selling of English suburbia in the interwar period, where the promotional message contrasted sharply with prejudices against suburbia endemic in intellectual circles. The same theme emerged when considering representations of industrial cities, where the focus switched from emphasising traditional advantages for manufacturing industry to rebranding as cities of culture. The case study of Detroit in particular brought home the difficulty of creating new identities for places that are inclusive, rather than constructed to meet the interests of those with power. These themes are taken further in the next chapter where we explore environmental representations concerning historic and future cities.

Further reading

Three valuable syntheses that touch on approaches to representations of the city are:

Ash Amin and Nigel Thrift (2002) *Cities: reimagining the urban*, Cambridge: Polity Press.
Mark Gottdiener (1995) *Postmodern Semiotics: material culture and the forms of postmodern life*, Oxford: Blackwell.
John Rennie Short (1996) *The Urban Order: an introduction to cities, culture and power*, Oxford: Blackwell.

On the question of endemic values towards the city, see:

Alistair Clayre, ed. (1977) *Nature and Industrialization*, Oxford: Oxford University Press,
Andrew Lees (1985) *Cities Perceived: urban society in European and American thought, 1820–1940*, Manchester: Manchester University Press.

On neighbourhood life and security, see:

Barbara Eckstein and James A. Throgmorton, eds (2003) *Story and Sustainability: planning, practice and possibility for American cities*, Cambridge, MA: MIT Press.
John R. Gold and G. Revill, eds (2000) *Landscapes of Defence*, Harlow: Prentice Hall.

With regard to place promotion and its role in policies concerned with reimagining the city, see:

John R. Gold and S.V. Ward, eds (1994) *Place Promotion: the use of publicity and public relations to sell towns and regions*, Chichester: John Wiley.
Tim Hall and Phil Hubbard, eds (1998) *The Entrepreneurial City: geographies of politics, regime and representation*, Chichester: John Wiley.
Stephen V. Ward (1998) *Selling Places*, London: E. and F.N. Spon.

8 Historic cities, future cities

This chapter extends the analysis of representations of the city by considering:

- representations of historic cities, with particular reference to landscapes of power and the narratives associated with heritage;
- the role of utopianism and dystopianism in representations of the future city;
- representations of the future city that exemplify these positions, contrasting the very different portrayals found in science-fiction film and the literature on urban sustainability.

Landscapes of power

Scholars often compare cities to text (Short, 1996, 390). A favoured metaphor likens the existing city to a *palimpsest*, the name given a piece of parchment or other writing-material written upon twice, on which the original script has been erased to make way for the second. This occurred because parchment, in particular, was expensive and capable of re-use, but the process of erasure was often incomplete. Close examination of the parchment allows the trained eye to discern at least traces of the original writing. Applying that idea to cities, urbanists have shown that it is possible to trace previous stages in the city's development despite the overlayering of the present. Another popular, and related, metaphor envisages the city as a *book*. Kevin Lynch (1960), for instance, wrote about the city's 'legibility', which he defined as the ease with which individuals can organise the various elements of urban form into coherent mental 'images'. Writing from an architect's perspective, he argued that cities varied in the extent to which they evoke a strong image – a quality he termed 'imageability'. Lynch argued that 'imageable' cities were places that could be apprehended as patterns of high continuity with interconnected parts. In other words, a city was likely to be 'imageable' if it was also 'legible'. On a related theme, David Harvey (2001, 128) reported that: 'A city centre . . . is a great book of time and history'. Its buildings, individually and in groupings, contain important information as to what the city is currently about and what it has been. What is necessary is to learn how to read the signs and to incorporate them into a convincing explanatory narrative. One area in which these skills are invaluable is in interpreting the way that powerful groups in society have moulded the urban landscape of their day.

Representation and power, as we have already suggested, are interrelated. In cities, buildings, statuary, memorials, street design, provision of open spaces and skylines all provide convenient media by which the rich and powerful can express ideas about themselves and their status within society. The ruling dynasties of medieval Europe, for example, created ceremonial landscapes of squares, triumphal arches and avenues for the staging of spectacular processions, tournaments, state entries, fetes, masques and masquerades. Such spectacles were essential vehicles in a ritualistic and symbolic society, surrounding rulers with the mystical aura of authority that helped legitimise their power over their subjects (Strong, 1984, 40). These features often remain as a remnant of past relationships between rulers and ruled from a past era. The great cathedrals and palaces in the major cities of the former Austro-Hungarian Empire or the Italian city-states speak of their past role as major centres of spiritual and temporal power. The neo-Classical colonnaded banks, corporate headquarters, museums, libraries and civic buildings of the nineteenth-century city clearly articulated ideas central to the development of power in the Victorian era. Equally the accentuated skyline of the modern financial district, with its landmark buildings of metal, glass and reinforced concrete again express landscapes of power. Each period and each gathering of capital, then, leaves its traces in the built environment of the city through such media as:

- styles of architecture;
- conspicuous display of wealth;
- constructing memorials;
- building walls and barriers;
- providing forms of surveillance, including security cameras;
- appropriating areas of urban space for private use;
- creating symbols of ownership or authority.

The English city of Bath, one of Britain's oldest cities, exemplifies these points. Situated on a bend of the River Avon in a bowl of the Cotswold Hills, it has a continuous history comfortably predating the Roman invasion of AD 43 (Cunliffe,

1971). The Roman town, like the earlier Celtic settlement, centred on the thermal springs. Although declining markedly at the end of the Roman period, during the Middle Ages Bath regenerated into an important trading and ecclesiastical centre protected by a town wall and dominated by the temporal and spiritual power of the Church. After the Church's power was severely curtailed by the dissolution of the monasteries in 1539 and the surrender of church lands, the centre of power moved to the municipality. During the eighteenth century, Bath entered what many regard as its second 'golden age'. Favoured by Royalty, it became a leading spa town, with some of the best examples of what is now termed Georgian architecture. Although this period still defines the architectural character of the city, more recent development introduced further changes. The Victorian period, for example, witnessed the arrival of the railways and the development of small factories and warehouses. The mid-twentieth century saw the introduction of radically new ways of handling the physical fabric of the city and traffic flows. In each period, these changes can be related to the ways that ruling groups found ways to express their position by leaving the physical impress of their culture on the city. Exercise 8.1 supplies more understanding of this process.

Exercise 8.1

Read through the following case study, which provides brief portraits of the city of Bath at four key times in its history – respectively, Roman (AD 43–410), Medieval (750–1500), Georgian (1714–1830) and modern (1950–75). As you go through, for each period make a list of:

- Who had power?
- How was that power expressed in the landscape of the city?

Beginning with the Romans, what is now presented as *Roman Bath* is primarily archaeological restoration, but a surprising amount is left of the original spa. Indeed with the exception of Hadrian's Wall in the Anglo-Scottish Border Country, the Roman remains at Bath are regarded as 'the best preserved, most famous and most impressive monuments of the Roman era to be found in Britain . . . among the most remarkable remains of this kind north of the Alps, and may be said to rival in some sense even the great imperial structures of Rome itself' (Pevsner, 1958, 97). At first the Roman settlement of Aquae Sulis (or 'the waters of Sulis') had military significance, but quickly developed as a resort and religious site around the hot springs. Between AD 65 and 75, the Roman administration built an extensive temple and spa complex dedicated to the goddess Sulis Minerva. This was a conflation of the Celtic dedication of the spring to the goddess Sulis with the Roman goddess Minerva – a synthesis of the two mythologies that conveniently preserved the link with the pre-existing culture while culturally assimilating the site to Rome (paralleling the subsequent policy of the Christian Church, see Chapter 4). The town developed within a 22-acre walled enclosure that housed the Romano-British elite, beyond which lay considerable amounts of extra-mural housing. The baths also became enclosed spaces. Originally open to the elements, by the fourth century they were covered by vaulted roofs. They catered for the local citizenry and notable visitors, who came to take the waters or seek medical cures, but excluded others. Characteristically the Victorian renovators placed a set of statues of Roman dignitaries, sculpted by George Lawson in 1894, around the walls of the Great Bath. Quite apart from the aesthetic sensitivities of the time, this was an irresistible opportunity to use statuary as media to make links between Britain's Roman past and the perceived glories of its current Empire (Figure 8.1; see also Chapter 6).

Figure 8.1 *Bath Abbey and Baths*

The locus of power in *Medieval Bath* remained adjacent to the heart of the Roman city. The settlement initially decayed after the Roman legions left in the early fifth century AD, with the roofs of the baths collapsing and the population sharply declining. Bath gradually regenerated into a walled settlement during the Medieval period, with its development connected with the complex history of the local Christian Church and changing church-state politics. The present Abbey (Figure 8.1), founded in 1499 and situated on part of what had been the Roman temple precinct, was heir to at least three previous churches and possibly a small Roman temple (the *tholos*). A Saxon monastery church, built in 781, succeeded a nunnery founded in 676. When the last abbot of Bath died in 1087, the abbey and its lands were granted to John de Villula from Tours in France (also known as John of Tours), Bishop of Wells. This development concerned more than just ecclesiastical administration. John combined the role of spiritual head and principal landowner, a merger of spiritual and temporal that continued for over four centuries. The church itself was rebuilt in 1106, destroyed by fire in 1137, rebuilt on altogether grander lines, but in a ruinous state when replaced by a smaller, nevertheless magnificent Abbey.

This brief historical sketch shows that the Roman Catholic Church and its powerful Benedictine monastic order originally owned not only the land immediately occupied by the church but was also the town's main landowner. The adjacent marketplace, controlled by the Church, specialised in the wool trade – as did the monks working in the monastery. Little remains of Medieval Bath due to its obliteration by the Georgians (see below), but the spiritual and temporal power then wielded by the Church is still amply visible in the scale and lavishness of the Abbey and the opulence of its fitments. Perhaps its most famous feature is the one that underlines the power of Christian belief, and theology, as arbiter of the individual's salvation. Buttresses on the West Front contain stone ladders to heaven on which miniature angels ascend and descend (Figure 8.2). Some, literally fallen angels, tumble as they climb. Facing out from the Abbey's main entrance, the angelic ladders conveyed important symbolic messages to the onlooker in this, largely pre-literate era (see also Chapter 4).

Georgian Bath is often incorrectly described as a planned city, but actually developed piecemeal during the eighteenth century through speculative ventures by the Corporation and by property developers. Its apparent cohesion came less from overall notions of planning than from common use of the honey-coloured Bath stone and common reference to late-Renaissance or Palladian Classicism as an architectural vocabulary (Jackson, 1991, 5). These developments met the demands of the increasingly wealthy and more self-confident aristocracy, then becoming independent of the rule of both the monarchy and the church and playing a growing role in the government of the country. Those who came tended to divide their year between time spent at Bath, with that spent on their country estates or at their town houses in London. Their activities focused on pleasure, variously taking the waters, socialising, dancing, gambling, attending the theatre and listening to concerts.

Figure 8.2 *Angelic host, Bath Abbey*

Bath's new elegance sustained a prolonged building boom. Although it barely exceeded the size of the Medieval settlement in 1750, Georgian developments propelled Bath from Cotswold market town to city with an international reputation by 1800 (Borsay, 2000, 11–16). Key developments within the historic core included new civic buildings and places of entertainment. Outside the core, developers built extensive new estates on vacant land, especially to the north and west. These developments contributed many of Bath's showpiece buildings and estates: notably, the Circus (1754–8), Royal Crescent (1767–74), the Rialto-inspired Pulteney Bridge (1770) and the Pump Room (1789–99). Those who attended for the season created demands for new facilities: the theatre, assembly rooms, hotels, tearooms, concert halls and bathing facilities. At the same time, they undeniably left the impress of their tastes, through their wealth and power, on the emerging city.

The preference for neo-Classicism was one such expression. The nobility drew parallels between their own lives and those of the learned aristocracies of classical civilisations in Greece and Rome – preferences reflected in their tastes for architecture, art and literature. In terms of the built environment, it meant adopting forms and façades of Palladian neo-Classicism, especially through the work of

John Wood and his son (also named John). Architects and builders added colonnades, friezes, statuary and other devices to communicate messages about the civility and sophistication of the residents. The use of geometric shapes – crescents, squares, circuses and the like – imposed the builders' and occupants' will on the landscape in a way that contrasted markedly with the organicism of the Medieval city. They also clearly demarcated the Georgian social order in the landscape. Figure 8.3, for example, shows a haha used to demarcate the private lawns of the Royal Crescent from the parkland beyond, in precisely the same manner as for a grand country house. By this means, the residents gained a sweeping view of green space without the visual interruption of a fence.

Figure 8.3 *The Royal Crescent, private gardens and haha, Bath*

The architecture had an explicit moral dimension. The vernacular styles found in Medieval Bath were now regarded not just as outmoded, but also failing to create continuity with the greatness of Bath's past and all that it stood for. These attitudes helped to generate the sack of the Medieval city in the eighteenth century. Using private Parliamentary Acts and other devices, the city authorities either demolished all the buildings of Medieval Bath or rebuilt them so that the street facade at least was in the approved style. This included rebuilding the parish churches, destroying the Abbey House (1755) and the early Stuart Guildhall, and removing the remaining town gates and fortifications (Borsay, 2000, 145–6). Medieval Bath then largely ceased to exist, apart from the Abbey and certain elements of the street pattern (Fergusson, 1973, 11). These actions carried enormous implications for future perceptions of the city. By 'obliterating the spa's historic landscape, the Georgians effectively destroyed the opposition. When future generations turned to the traditional townscape as a source of pleasure and profit, there was in Bath – the Abbey and Roman baths excepted – only one pre-Victorian heritage to be recovered, that of the Georgians themselves' (Borsay, 2000, 147).

The reshaping of *Modern Bath*, therefore, had a precedent. Between 1955–72, a significant part of Bath's political establishment believed in the need to reconstruct in the interests of modernisation and progress. Like other historic cities, Bath struggled to accommodate the demands of the present era alongside a venerable, and protected, historic fabric and a street pattern dating from a pre-motorised age. The centre faced traffic problems and a lack of modern retailing space. Areas of working-class housing in the south of the city, as well as some architecturally significant portions of the Georgian city, had fallen into disrepair through neglect. New housing, civic buildings and hotels were required. The spirit of the age favoured comprehensive solutions to problems and empowered the local authority's architects and planners to deliver those solutions. While requiring new buildings to preserve the look of the city by maintaining the building line and having their street façades faced with Bath stone, their training predisposed them to apply rational principles of design, to favour technologically sophisticated techniques

Figure 8.4 *City of Bath Technical College*

and to advocate comprehensive rather than piecemeal solutions to endemic problems. They were assisted in enacting those preferences by a supportive, even enthusiastic, social consensus. It was firmly believed that the specialist knowledge of social scientists and designers could produce a better world – where 'better' meant cleaner, safer, more efficient and, often, more socially just. The new period, therefore, brought new ideas about design, aesthetics and planning.

In Bath, the burden of clearance fell on the artisan quarters of the city – the terraced houses of those who serviced the Georgian city. As the drive to redevelop quickened, large-scale development occurred in an attempt to clear the ground and modernise the city, primarily concentrated in five areas around the core. In its way, such clearance was unexceptional by the standards of 1960s Britain and, indeed, Bath Corporation took active steps to restore and enhance aspects of the city's historic fabric. Nevertheless, the architectural jewels of the Classical inheritance were left intact at the expense of their context. The less valued elements of the historic townscape were demolished to make way for new developments. In the south of the city, for example, planned clearance swept away extensive areas between the Abbey Green and Bath Spa railway station to make way for new traffic schemes, including pedestrianisation and traffic diversions, retail developments, a new eight-storey Technical College (Figure 8.4), medium-rise flats, a bus station, and a hotel.

Table 8.1 *Landscapes of power in Bath*

Period	Groups exercising power	Landscapes of power
Roman (43–410)	Romano-British ruling class	Appropriation of Celto-British religious site; construction of spa and symbolic spaces of temple complex; walled enclosures; town walls and gates
Medieval (750–1500)	Roman Catholic church, monastic order, local dignitaries	Development, aggrandisement and embellishment of church; control of adjacent marketplace; town walls and gates
Georgian (1714–1830)	City corporation, landowners, architects, speculative builders, aristocracy	Destruction of earlier buildings and fortifications; imposition of new aesthetics, e.g. late Renaissance classicism and rigid geometric order;

Table 8.1 *continued*

Period	Groups exercising power	Landscapes of power
		re-establishment of link to Roman legacy
Modern (1950–75)	City corporation, municipal planners, civil engineers and architects	Socially oriented town planning; comprehensively redeveloped areas; new building materials and aesthetics

Looking through this evidence, your answers to the exercise might well contain elements shown in Table 8.1. The exercise selected just four periods in Bath's history but showed how, in each, different ruling elites imposed their sense of order, wealth and culture on the city's landscapes. These snapshots of Bath at four different periods in its history, then, are at once historic and contemporary. They are historic in that they point to the way that elite groups in the past represented their power by shaping the city of their times. The power and authority of Rome, exercised by its Romano-British administrators and priestly class, was replaced by the temporal and spiritual power of the Christian Church during the Middle Ages. Their power as landowners was broken in the sixteenth century through the surrender of lands to the Crown and the Dissolution of the Monasteries. Power subsequently passed to the city's Corporation, with Bath growing rapidly during early modern times under the Georgian administration. We finally saw it 'modernised' in the period after the Second World War by the city councils and their professional built environment specialists. Yet the story did not end there. The rise of the modern was itself short-lived. Within 15 years, the 'ascendant ethos' of town planning that had accompanied it was under attack nationally as well as locally (Borsay, 2000, 182). The driving force that fundamentally altered much of the area around Bath's city core foundered. The rhetoric of reconstruction based on new ideas of form and function ceased to attract adherents. The clearance machine halted, due partly to the perceived indifferent results of redevelopment and partly to growing restrictions on the public purse. The rise of the conservation lobby and concern with heritage also revalued that which had received little protection during the 1960s, culminating in UNESCO designating the city of Bath *as a whole* as a World Heritage Site in 1987. Conservation plaques, literature and new interpretative centres codify new ways of looking at the city and portraying its landscapes. The familiar process of constructing and reconstructing environmental representations continues.

Narratives of heritage

The question of 'heritage', however, merits further discussion. It is a complex term, associated with inheritance and the idea of property (lands, sites, landscapes, buildings, artefacts, ideas, customs) being passed on from one generation to the next. As such, *heritage* essentially comprises the valued legacy of previous generations; items from the past that embody tradition and which, by current evaluation, are considered worth retaining for the benefit of present and future generations. Some also argue that, as the past-in-the-present, they are frequently important sources of national, local, even individual identity (e.g. Hewison, 1987, 15–17). 'Ownership' of that property is often contested, for inheritance is sometimes associated with disinheritance and possession with dispossession. Just as potential inheritors can contest the disposition of the deceased's property under the terms of a will, so can groups in society when confronted with dispossession from lands, sites, buildings and memorials to which they feel they have a legal, moral or spiritual claim. In addition, what constitutes 'heritage' is a selective and changing notion. Those who have power repeatedly make choices over how to value the past and what to do about it. The next exercise helps to elaborate this point.

Exercise 8.2

The three extracts that follow deal with the same place, the cave dwellings in the town of Matera in the Basilicata region of southern Italy. The extracts date from different times, respectively, the mid-1930s, 1955 and 2003. After reading through, list the differences that you find.

Why do you think these representations of Matera's cave dwellings differ so much?

In the gully lay Matera . . . The narrow path wound its way down and around, passing over the roofs of the houses, if houses they could be called. They were caves, dug into the hardened clay walls of the gully, each with its own façade, some of which were quite handsome, with eighteenth-century ornamentation. . . . The houses were open on account of the heat, and as I went by I could see into the caves, whose only light came in through the front doors. Some of them had no entrance but a trapdoor and a ladder. In these dark holes with walls cut out of the earth I saw a few pieces of miserable furniture, beds and some ragged clothes hanging up to dry. On the floor lay dogs, sheep, goats and pigs. Most families have just one cave to live in and there they sleep all together; men, women, children and animals. This is how twenty thousand people live.

Of children I saw an infinite number. They appeared from everywhere, in the dust and the heat, amid the flies, stark naked or clothed in rags; I have never in all my life seen such a picture of poverty.

(Levi, 1948, 86–7)

The notorious cave dwellings of the town of Matera are also found in the villages sited in the limestone canyons on S. border of the Murge. The town is sited on the flat plateau top, the cave dwellings are cut into the steep limestone bluffs of the valley and over 10,000 persons live in them. New villages are being built by special act of Parliament to evacuate these people, who often trek up to 10 km. daily to and from their fields.

(Dickinson, 1955)

Discover . . . I Sassi di Matera – Basilicata
Matera: a city unlike any other in the world, carved out of white tuff and the blinding Mediterranean sun. . . . The Matera of the 'Sassi' is a city carved out of the rock, formed by one of the most suggestive city environments in the world, classified by Unesco as a legacy of humanity to be passed on to future generations. The Sassi, Barisano and Caveoso, set on the deep slopes of a valley, with cavernous meanders and underground labyrinths, hide relics of a remote past rich in culture and history. Caves and underground architectures are connected by steps carved out of the cliffs and are enclosed by dry-stone walls bounding small vegetable gardens, authentic stone gardens to be tilled, where the fecund strength of the vegetable world reveals its utmost potency. Wandering along the streets of this underground city you run into ancient places of worship, used to celebrate in the depths of the earth the nuptials of sun and stone.

(Italian Tourism, 2003)

These three quotations identify profound changes in representing the same place over a period of around 65 years.

1   The first comes from Carlo Levi's *Christ Stopped at Eboli* (1947), an account written about the time in 1935–6 that the author spent in political exile in a remote part of Basilicata. In the book, Levi recounts his sister's description of the scene at Matera where she had gone to get authorisation to visit him at Aliano. She recognised the spectacular nature of the setting and the fine ornamentation on the façades of some dwellings, but recoiled at the appalling living conditions and the poverty with an intensity reminiscent of the mid-Victorian novelists' writings about the industrial city. Given that she was a doctor, Levi's sister recognised the symptoms of malaria, trachoma, malnutrition, lice and scabies, especially amongst the children. Levi structured this account to telling effect, alerting the reader to the injustice suffered by the people in a neglected part of the Italian state during the Fascist era. Little is found to balance the horrors that the visitor discovers and nothing is done to suggest that the future is anything other than hopeless.

2   The second quotation, from the work of the geographer Robert Dickinson, stems from a period in which planners were making a contribution to tackling previously intractable problems. The days of the 'notorious' cave dwellings were numbered (they were condemned as unfit for human habitation in 1952). A special Parliamentary Act would provide new villages for the cave dwellers and allow for the evacuation of their residents. The planned settlements would also improve economic efficiency, reducing the lengthy treks between the people's homes and their fields. Progress has been achieved and a better future was implicitly at hand.

3   The third extract shows Matera now firmly on the tourist trail. Recognised by UNESCO as a World Heritage Site in 1993, the cave dwelling or i Sassi (literally 'the stones') were labelled as 'one of the most suggestive city environments in the world'. The abandoned cave dwellings had become a development opportunity. What were once dreadful, disease-ridden hovels were repackaged as heritage. The deficiencies of the past were a legacy that constitutes economic resources for the future. The visitor was invited to wander in this hidden and remote place, discover the richness of its treasures, and experience the spirituality of its ancient places of worship. The subject of Carlo Levi's powerful evocation of human misery had become a place for the visitor's enlightenment, even self-discovery.

Collectively, these quotations indicate a remarkable change. Abandoned buildings in a remote southern Italy town now feature on UNESCO's list of World Heritage Sites alongside Stonehenge, the Sphinx, Ankhor Wat, the Acropolis and, indeed, Bath. The Sassi are not only protected, they are now being actively conserved to develop their heritage significance: their UNESCO citation declaring that it was 'the most outstanding, intact example of a troglodyte settlement in the Mediterranean region, perfectly adapted to its terrain and ecosystem' (UNESCO, 2003). One former dwelling, the Cave Dwelling of Vico Solitario, has become a heritage centre. Its publicity reassuringly characterises the site as an '*ancient* dwelling with furnishings typical' (emphasis added), obscuring the fact that these dwellings were fully occupied well within living memory (Comune di Matera, 2003). The invitation to the potential visitor is couched in terms of this attraction showing local customs, the organisation of family life, and the fitments and utensils associated with these dwellings:

> In order to gain a better understanding of the customs of the inhabitants of the 'Sassi', the old town of Matera, before it was abandoned, you are invited to visit a peasant dwelling.

> The habitant located in Vico Solitario near the Church of San Pietro Caveoso is a typical cave dwelling with furniture and tools of the time so as to give a visitor an exact idea of how family life was organized in the households of the 'Sassi'.

Without having to act as overcrowded living space, with all the attendant problems of hygiene and foul air, the visitor sees a museum that sanitises the experience of this place. Elsewhere in the 'abandoned' town, developers convert former cave dwellings to 'sympathetic uses', such as shops, hostelries, restaurants and souvenir shops. Some are even being redeveloped as dwellings – described as 'trendy digs (for) artistic types' (Gogermany, 2003). The more recent past, though, is not totally forgotten. Bearing in mind that some tourists will want to 'experience Carlo Levi's Matera' after having read *Christ Stopped at Eboli*, they are reminded to visit the house where he lived at Aliano, which is itself open as a museum.

The recycling of Matera's poverty and deprivation into themed heritage is unexceptional by national or international standards. Like the rise of conservation, with which it is indissolubly linked, heritage interpretation is a global phenomenon (Lowenthal, 1996). Originally a term which embodied revelation and provocation, 'heritage interpretation' is an activity which presents, structures and reflects on information in order to help the visitor to make sense of the places or objects in question. As part of the strategic functions involved in the management of heritage, it is closely linked with communication and, increasingly now, with promotion and marketing. Goodey (1994, quoted in Harrison, 1994, 312) suggested that interpretation had four essential elements:

- understanding the nature of the audience (the market);
- identifying the themes and stories to be told;
- identifying the resources to be used in interpreting these themes and stories, and
- considering the most appropriate and effective media to be used in the context of the three previous criteria.

While this analysis appears logical and systematic, the stories or narratives told by interpretive activity depend very much on the ideology of the storyteller. Consciously or otherwise, interpretation compartmentalises knowledge, directing the observer to identify specific elements or relationships and to neglect others, and argues fiercely for the retention of some things while allowing others to be demolished. In the early 1960s, city authorities throughout the Western world demolished classic examples of nineteenth-century industrial and commercial buildings, as well as swathes of dilapidated, but structurally sound housing. Had the same buildings met an identical threat 20 years later, the outcome would probably have been quite different.

Two important considerations flow from this analysis. First, the changing narratives that underpin heritage interpretation clearly have relevance beyond mere academic curiosity. Representations of the city's past are deeply implicated in emerging policy. Second, the past is not an unchanging or independent object, but something that is always revised from present positions (Crang, 1994). There is no such thing as a single 'authentic' history that can be set against other 'ideological' histories. Rather, interpretation is inherently a realm of multiple meanings and multiple explanations of those meanings.

Cities to come

So far in this chapter, we have linked changing representations of the city, and the values that underpin those representations, to questions relating to the historic city. Further important issues arise, however, when looking at representations of the future city. Cities are integral to most visions of the future; already dominating economic and cultural life in advanced nations and increasingly doing so in developing countries. Most forecasters choose to present their visions of the future in a studiedly neutral manner, but in reality the nature of their assumptions often predispose them towards what are known as either dystopian or utopian views of the future city. Before looking at examples of thinking about the prospective city, therefore, it is worth considering the precise meanings and implications of these terms.

Exercise 8.3

The word 'utopian' is often used casually when thinking about schemes for the future. List any features that you would expect a utopian project to have.

Now read through the following definition. When you have done so, add any other features that might be missing from your list.

Utopia and dystopia

We have referred continually to the polarity of thought continued in representations of the city down through the ages. The dialectical character of the thinking about the city, however, often meant more than just seeing the city as 'good' or 'bad' on a priori grounds. Rather, observers of the city came to believe that the inherent qualities of the city themselves exerted a transforming influence on those who lived there. For some, like Plato and the Socratic philosophers, the city was a crucible for social progress and the achievement of the Good Life. For others, the city was a breeding ground for temptation and all forms of evil that would lead people into sin, degradation and misery. In time, these positions would respectively come to be termed 'utopian' and 'dystopian'.

The term 'utopian' comes from Sir Thomas More's monograph *Utopia*, written in 1516. Ostensibly the book is a travelogue about an island kingdom situated somewhere in the Caribbean and based on the recollections of the wandering philosopher Raphael Hythloday. As such, its account of exotic lands and their customs would have been no stranger than many other reports circulating in sixteenth-century Europe. The narrator claimed to have found an ideal land of justice and happiness, as the book's Prologue apparently makes clear (More, [1516] 1965, 21):

> The ancients called me Utopia or Nowhere because of my isolation. At present, however, I am a rival of Plato's republic, perhaps even a victor over it. The reason is that what he delimited in words I alone have exhibited in men and resources and laws of surpassing excellence. Deservedly ought I to be called by the name of Eutopia or Happy Land.

The book's coded character was apparent from the outset. Like many works subsequently labelled 'utopian', it appeared at a turbulent time in world history. The first phase of the Renaissance was over and the Reformation about to begin. Contemporary England contained many internal political, religious, economic and social tensions. As such, the book's form was a safer way for an influential person – More was later Lord Chancellor of England – to express dissatisfaction with the status quo than publishing openly dissenting treatises. More coined the word 'utopia' as a deliberate pun on the two Greek words that supply the 'u' sound: *eu* (good) and *ou* (not). When taken together with *topos* (place) – the root of the second part of the word – utopia instantly contained ambiguity. It could be interpreted as either the 'good place' or the 'not-place', the place that does not exist. In creating this play on words, More not only supplied a generic term for an incredibly varied species but also encapsulated the point that has haunted so much discussion: namely, is utopia the search for the best, for the ideal, or is it a fruitless quest for the impossible? Often utopianism is 'an invitation to perceive the distance between things as they are and as they should be' (Eliav-Feldon, 1982, 1) and, as such, is as valuable for its critical commentary as for its conceptions of a better world.

As More's reference to Plato made clear, he was not the first to attempt a comprehensive vision of an ideal society. Nevertheless, he produced a powerful new synthesis that bound together the two ancient belief systems that had nurtured and moulded utopia: the Hellenic myth of the man-made ideal *polis* and

continued

the Judeo-Christian faith in a paradise created with the world and destined to outlast it (Manuel and Manuel, 1979, 16–17). While inherent contradictions arise from so rationalising the earthly and celestial cities, More successfully re-established the centrality of the city in utopian thought. As such, cities have frequently figured in schemes for utopia, with architects, planners, philosophers, political theorists and others designing cities as crucibles for social transformation (see Tod, 1982; Fishman, 1977).

Others, however, are less convinced. They see utopianism as a dangerous diversion; an exercise in wish fulfilment that sets aside real world constraints in pursuit of the author's dreams. They argue that most utopians have a serious preference for their proposals and for the notion of the Good Life that they embody (Goodwin, 1978, 4), but highlight the fact that they rarely have any clear idea of how the new world is to be brought about. Usually the only way utopias can make the transition from *here* to *there* is by means of authoritarian power backed by what David Harvey (2000, 171) termed 'a restrictive set of social processes'. Utopians may argue that this will be benign, enlightened and justifiable given the ends that they have in mind, but their critics disagree. The distinguished American urbanist Lewis Mumford (1965, 278) once asked: 'How could the human imagination, supposedly liberated from the constraints of actual life, be so impoverished? [.] where did all the compulsion and regimentation that mark these supposedly ideal commonwealths come from?'

Critics also observe that creation of utopia always invokes its antithesis, 'dystopia'. *Dystopia* is the name given to an imaginary place or condition in which everything is as bad as possible. The word itself has a much shorter history than 'utopia'. Although first used by John Stuart Mill in a Parliamentary speech in 1868 as an opposite to utopia, its modern usage only dates from the 1950s (Negley and Patrick, 1952, 298). Nevertheless, the ideas to which it relates are equally old, since the dream image of Heaven is incomplete without the nightmare vision of Hell. The ordered tranquillity of the former contrasts with the disturbing chaos of the latter. Dystopians agree with utopians in maintaining that the city is indeed a crucible for social transformation, but they see the transformation as wholly negative. From their standpoint, the city variously impoverishes human life, crushes individualism, alienates and encourages oppression.

From this analysis, if something is referred to as 'utopian', it can mean any of the following:

●   a state of affairs that represents the best possible situation;

- a vehicle for achieving the necessary social transformation that will deliver the Good Life;
- something that represents the search for the impossible, for that which does not exist;
- a standard or model that can be compared with the existing situation to show the deficiencies of the present;
- a state of affairs that is a recipe for authoritarianism;
- something that is incomplete without its opposite. Just as heaven is incomplete without hell, so the utopian is incomplete without the dystopian – the worst state of affairs.

This degree of ambiguity and complexity is always present when considering utopianism, as the next two sections clearly show. The first considers dystopian representations of the future city in the shape of the cities portrayed in science-fiction film. The second then examines utopian representations of the future city, as found in literature on sustainable cities.

The shadow of despair

Science-fiction filmmakers have had a deep interest in the city that dates from shortly after the First World War. The harnessing of technology for total war showed only too clearly the disturbing directions in which human society might develop. Early experimental films such as Hans Werckmeister's *Algol* (1920) and *Aelita, Queen of Mars* (1924) brought to life a pattern of expectation that was also then emerging in American science-fiction comics, which saw society's future allied to giant cities, particularly ones with a pronounced vertical dimension. They also clearly signalled the creation of a new bastion of the discourse of anti-urbanism. The explicit belief that such cities might well be inherent arenas of evil and repression was developed and codified by Friedrich ('Fritz') Lang's *Metropolis* (1926). Made in Berlin, *Metropolis* drew on influences that included German Expressionism (a contemporary modernist movement that strove to give artistic shape to inner states of mind), the powerful anti-urban rhetoric of Oswald Spengler's book *The Decline of the West* (1926), and a growing belief that 1920s New York was the face of a new, vertical city of the future. To Lang (quoted in Ott, 1979, 27), New York was:

> the crossroads of multiple and confused human forces (irresistibly driven) to exploit each other and thus living in perpetual anxiety. . . . The buildings seemed to be a vertical veil, shimmering, almost weightless, a luxurious cloth hung from the dark sky to dazzle, distract and hypnotise. At night the city did not give the impression of being alive; it lived as illusions lived. I knew that I had to make a film about all of these sensations.

In the event, the themes of exploitation and anxiety prevailed over celebrations of exciting urban forms. *Metropolis* was a film about class politics and oppression set in the year AD 2000. The city of Metropolis functioned as a complex three-dimensional entity (Neumann, 1999a, 34). Working upwards, its three levels consisted of the workers' city located deep underground; above them, but still subterranean, the machine halls and powerhouses that served Metropolis; and the surface level city of consumption. The surface city depicted soaring skyscrapers and high-level bridges towering over cavernous, restless traffic arteries.

It would prove influential. Although a handful of films emphasised the wonder and possibilities of such environments, the majority saw it as the recipe for dystopia. Further developments, such as the rise of the gangster movie in the 1930s and the grittiness and darkness of so-called *film noir* in the 1940s and 1950s, reinforced the cinema's negative portrayals of future cities. The machine-dominated environment of *Metropolis* recurred in several films (*Alphaville*, 1965; *Fahrenheit 451*, 1966; *Logan's Run*, 1976), in which the cities are technological marvels but where people live impoverished lives under strictly imposed authoritarian codes. New York lost credibility as the city of wonder, with its severe social and financial problems denting its ability to act as a credible setting for the 'effervescent high life depicted in celluloid skyscrapers' (Albrecht, 1999, 39).

The good city mainly persisted through the extraterrestrial thread of city-sized flying machines benignly ruled by Captain James Kirk and his ilk, but back on earth things looked bleaker (Gold, 2001). The quintessential science-fiction city was a dark and claustrophobic place: sometimes a city of perpetual night lit only artificially; sometimes one where the sombre skies constantly teemed acid rain; and frequently a city in which the air was heavily stained by industrial pollution. Stanley Kubrick's *A Clockwork Orange* (1971) set new standards for the determined way that its opening scenes created an expectation of urban violence. Richard Fleischer's film *Soylent Green* (1973), with its proto-environmental message, depicted a clammy and claustrophobic New York where the population had spiralled to 40 million by 2022. On a different tack, John Carpenter's *Escape from New York* (1981) contained a scenario by which Manhattan Island had become a maximum-security prison by 1997, hermetically sealed by a 50-foot high perimeter wall.

Nevertheless, if *Metropolis* is regarded as the work that crystallised the screen portrayal of the vertical city, then Ridley Scott's *Blade Runner* (1982) codified the darkness of the postmodern science-fiction city, usually referred to as 'future city noir'. The visual representations of the city now related to Los Angeles, but parodied or even inverted the popular myths that had helped promote Los Angeles in the 1970s – a paradise of booming aerospace industries, sprawling freeways, and leisure amid near-perfect weather (Albrecht, 1999, 42). Instead the Los Angeles of November 2019 was characterised by urban decay, deprivation, malfunctioning experiments in genetic engineering, and environmental pollution. It was an urban wasteland, where the only notable features were the buildings associated with the forces of control.

Although New York-inspired vertical city notions continued to have attraction, future city noir representations maintained their hold. The increasing output of science-fiction films, perhaps filling the narrative gap left by the decline of the cowboy Western, revealed a repetitive and clichéd visual content. Post-apocalyptic industrial chic rather than modernist sterility reigns. Skies remain darkened by rain or coloured by industrial pollution. The industrial wastelands offer a chaotic maze in which chase sequences take place. Probable and improbable motorbike-riding action heroes cope with the hazards posed by automated machinery and untended chemical processes, as well as the dangers of the formidable, but ineffectually used, firepower available to their opponents. The crumbling city remains a battleground, with any dark corner potentially containing an adversary. If any urban schema exists, it is the spiders' web, at the heart of which lies the high-rise, high-tech headquarters of whichever Evil Empire is currently being fought.

Having said this, a new prototype has recently emerged through the idea of virtu-ality. The ending of the Cold War has diminished the clarity of 'the enemy' in science fiction, with the intelligent machine now assuming the place of the authori-tarian regime as opponent for the righteous. The entrapment of human beings in virtual environments by intelligent, power-usurping machines has precedents, but the Wachowski Brothers' film *The Matrix* (1999) was the first to work this notion into a full scenario about the city. The key lay in the concept notion of *hyperreality*. Derived from the work of the French cultural theorist Jean Baudrillard (e.g. 1994; see Poster, 1998) and brilliantly developed by the Italian semiologist Umberto Eco in an essay written in 1975 (see Eco, 1986, 3–58), 'hyperreality' was the state in which the copy or 're-creation' becomes more convincing, and seemingly more authentic, than the real thing. In the case of *The Matrix*, the action featured the clash of two 'realities': one that was apparently the everyday world and another that lay behind it. The everyday world was eventually exposed as hyperreality constructed by the machines that controlled society and preyed on their human victims. The hyperreal virtual city conveniently replicated cities as they appeared in the late twentieth century. By contrast, the real world was characterised by a late twenty-first-century future city noir, a nightmare world in which the human rebels survived in hidden subterranean corners surrounded by industrial dereliction.

Not surprisingly in view of intellectual convention, science-fiction film emphasises the dystopian character of the city by accentuating its distance from nature and the repression of its inhabitants. A popular narrative theme was that of Escape to Eden by those fleeing from authoritarian regimes. An early example, *ZPG* (1971) was set in an overcrowded city permanently wreathed in sulphurous smog. A couple that had disobeyed the Zero Population Growth decree had been sentenced to death, but the new family escaped by raft along the main sewer. The green world beyond, with its promise of freedom and more humane values, offered them life and hope. The ending of the original theatre release of *Blade Runner* saw Deckard and the android ('replicant') Rachel escape from Los Angeles. Courtesy of out-take footage from another Warner Brothers' film *The Shining* (1980) shot at Moab (Utah), the audience saw bright sunlight and greenery in place of the gloom and perpetual rain of Los Angeles as the now-happy couple travelled confidently to their new life away from the grasp of the city. Quite how these environments have managed to remain so pristine while the city is so irrevocably ruined is never adequately addressed.

The celebrated opening sequences of *The Terminator* (1984), the original film in what is now a trilogy, provides an example in which city–nature relationship is used to telling effect in establishing the film's characterisation. *The Terminator* depicted the Los Angeles of the 1980s in which a battle of the *future* was being fought out. Briefly, a killing cyborg was sent by time travel back from 2010 with the mission of exterminating Sarah Connor, who would be the mother of a future resistance leader. By killing her, that leader (John Connor) would therefore never be born and the machines will be victorious. For their part, the resistance of the future dispatched a guerrilla fighter, Kyle Reese, to save Sarah from the Terminator. The film opened with the two protagonists arriving back separately from the future, with the action staged in environments that contrasted with their true identities. Both arrived at night and both were naked. The Terminator, played by Arnold Schwarzenegger, arrived first in a largely deserted suburban street. After forcibly taking clothes from three youths who chanced to be in the area, the scene ended with the android looking down from the hills over the romantic carpet of city lights shown below – the scene of his murderous quest. By contrast, Kyle Reese (Michael Biehn) arrived in an inner city street and, in searching for clothing, had to endure being chased though the maze of an unfamiliar city, coping with drunks, down-and-outs and siren-wailing squad cars filled with hostile police.

In a scene shortly afterwards, the Terminator made his way through leafy, sunlit suburbia to kill the first on a list of Sarah Connors that he has obtained by ripping a page out of the Los Angeles telephone book. The camera panned around a green neighbourhood, so safe that a small child was riding a tricycle unsupervised out-of-doors. The car driven by the Terminator then pulled up, with its front wheel neatly crushing a toy truck left in the gutter. This simple but effective use of place imagery underscored the narrative and characterisation. What made the crushing of the child's toy seem so shocking was that here was an Eden into which stepped a calculating android whose sole goal was to wipe out the innocent. It contrasted perfectly with the threatening city that Michael Biehn's character encountered, where a guardian angel, sent to save humanity, had to cope with an environment that pulsated with danger and evil.

Sustainable cities

If science fiction routinely deals with profound themes against the backdrop of a brooding, polluted and alienating city, others recognise the city has problems but believe that suitably enlightened radical action can bring the Good Life. Sometimes the chosen medium is benign application of appropriate technology. At other times, the emphasis is on voluntaristic or system-enforced behavioural change. In the interwar period, the chosen instruments of social transformation stemmed from enlightened architecture and planning. After blazing brightly in the two decades after the Second World War, those ideas foundered irretrievably in the light of experience of urban reconstruction. In the 1960s, freedom of mobility based on mass car ownership was the force seen as promoting new dispersed cities, functioning like a normal city but allowing people to live where they pleased. Those ideas broke down when the implications of increasing vehicle congestion and lengthening journeys-to-work became clear, namely, that the underlying vision was simply a greatly extended suburbia. In the 1980s, it was the turn of information technology. The combination of high-speed computing and low-cost telecommunications would supply the basis of 'wired cities' and 'electronic cottage societies', a new fusion of town and country that would combine the best qualities of both environments. The pattern was the same as before. Within several decades, the 'dot.com' crash and disillusionment with the social consequences of the adoption of information technology quelled the more extravagant hopes about the technology's ability to bring about the latest version of the 'brave new world'.

During the times when each of these sets of ideas about the prospective city flourished, their proponents would have happily accepted that their visions implied a better future while, no doubt, resisting any accusation that their visions were unrealistic or dangerously alluring. In each case, their representations were offered as applied commonsense, available to anyone who grasped the potential of new technologies and understandings. The lessons of hindsight, of course, are different. The applied commonsense is revealed as partial and flawed readings of the relationship between society and technology. Elements of the vision turn out to have been driven by the ideological interests of elite groups and the commercial corporations that stood to profit. The quick route to social transformation turned out to be as elusive as ever.

At the start of the twenty-first century the mantle of urban transformation – some would say the utopian mantle – has fallen to environmentalists and to designers of new 'green' cities. These designs normally attempt to apply the principle of 'sustainability' to the urban realm. Popularised by the Brundtland Report as meaning 'development that meets the needs of the present without compromising the ability of future generations to meet their own needs' (WCED, 1987, 43), sustainability is an essential part of modern environmentalism. How that notion is applied to cities, of course, depends on how it is interpreted. At one extreme, it need not mean more than managing cities to retain a constant stock of resources over time, finding substitutes where non-renewable resources are involved and setting aside income to pay for depreciation. Equally, more radical interpretations look for steps to achieve long-term commitment to multi-sectoral management of the urban environment, itself part of a programme of maintaining life on Earth in the most cost-effective and environmentally harmonious manner. Either position would involve changes in environmental ethics and values, but the latter implies far more in terms of socio-economic and political change.

Translating these ideas into principles for replanning cities was always likely to be one of environmentalism's greatest challenges. On the one hand, there is no doubt about the pressing need, with the twenty-first century as likely to be characterised by urban growth as was the twentieth. On the other hand, cities are not easily changed to meet the criteria for sustainability. They are major consumers of resources and producers of pollution and waste; roles which cannot be set aside lightly in the short term without severe economic dislocation. Small-scale and incremental changes to the status quo are unlikely to yield much in the short term and will inevitably frustrate those individuals and groups who want to move ahead quicker. The following exercise conveys something of the difficulties involved.

Exercise 8.4

The three extracts that follow represent recent capsule statements that seek to spell out the key principles behind urban sustainability. Read them and list:

a. the ways in which they differ

and

b. the common features that they show.

Now re-read the material earlier in this chapter on utopian thought. In what sense may these extracts be described as utopian?

Sustainable cities . . .

- have functioning infrastructures;
- need people with vision;
- are responsive to the needs of the people;
- are part of a wider sustainable context;
- address alternatives from a wider perspective;
- require active citizenship and good governance;
- have the capacity to identify problems and produce concrete solutions;
- listen to children, older people, minorities;
- deal with their garbage;
- recycle/regenerate themselves;
- are based on citizens' sense of ownership and responsibility;
- are convivial – enjoyable and fun places to live and visit.

(LSCP, 2003)

The sustainable city is:

- a *Just City*, where justice, food, shelter, education, health and hope are fairly distributed and where all people participate in government;
- a *Beautiful City*, where art, architecture, and landscape spark the imagination and move the spirit;
- a *Creative City*, where open-mindedness and experimentation mobilise the full potential of its human resources and allows a fast response to change;
- an *Ecological City*, which minimises its ecological impact, where landscape and built form are balanced and where buildings and infrastructures are safe and resource-efficient;
- a *City of Easy Contact and Mobility*, where information is exchanged both face-to-face and electronically;
- a *Compact and Polycentric City*, which protects the countryside, focuses and integrates communities within neighbourhoods and maximises proximity;
- a *Diverse City*, where a broad range of overlapping activities create animation, inspiration and foster a vital public life.

(Rogers, 1997, 169)

Sustainable communities are defined as towns and cities that have taken steps to remain healthy over the long term. Sustainable communities have a strong sense of place. They have a vision that is embraced and actively promoted by all of the key sectors of society, including businesses, disadvantaged groups, environmentalists, civic associations, government agencies, and religious organizations. They are places that build on their assets and dare to be innovative. These communities value healthy ecosystems, use resources efficiently, and actively seek to retain and enhance a locally based economy. There is a pervasive volunteer spirit that is rewarded by concrete results. Partnerships between and among government, the business sector, and nonprofit

> organizations are common. Public debate in these communities is engaging, inclusive, and constructive. Unlike traditional community development approaches, sustainability strategies emphasize: the whole community (instead of just disadvantaged neighbourhoods); ecosystem protection; meaningful and broad-based citizen participation; and economic self-reliance.
>
> (ISC, 2003)

These extracts appear in increasing order of the specificity of their recommendations. The first extract came from the website of the Learning for Sustainable Cities Project, which comprises a collaborative educational project based on six different cities: Manchester (UK), Banjul (Gambia), Brescia (Italy), Curitiba (Brazil), Halifax (Canada) and Mumbai (India). Perhaps understandably for a project that spans the interests of three cities in the developed world and three cities in the developing world, its proposals were extremely broad. It essentially argued that environmental actions must bear in mind the interests of all who are part of the community now, people in the wider world, and in generations to come. It supplied sets of broad conditions, without going into detail about the type of habitat that might result. Indeed, apart from using the word 'sustainable' once, there was little in this extract to distinguish a 'sustainable city' from other, previous types of ideal city schemes.

The next extract is quite different, although also using bullet points. Written by Richard Rogers, a leading British modern architect, it focused on the type of city as well as the broad conditions of sustainability. While dealing with ecological matters and access to resources, the extract bore the ideological hallmarks of architectural modernism. The sustainable city will value art, architecture and (designed) landscape, which are seen to 'spark the imagination and move the spirit'. It will value creativity, experimentation and open-mindedness. These are treasured elements in the lexicon of modernism – a school of architectural thought that internalised the expectation of rejection by 'conservative' forces in society. Rogers made suggestions for the shape of the city, but curiously advocates both compactness and the dispersal inherent in polycentrism. The new society would have community-based neighbourhoods but would function effectively over space.

The third extract, produced by the Vermont-based Institute for Sustainable Communities (ISC), changes the terms of the debate. The ISC is an American charitable foundation, originally established in 1991 in the wake of the collapse of communist rule in Eastern Europe, which attached its aim of assisting 'existing and emerging democracies' to the language of sustainable cities. However, rather than speaking directly about sustainable cities, it talked about sustainable communities, defined as 'towns and cities that have taken steps to remain healthy over the long term'. The authors set out moral standards and potential visions for all sectors of society that had strong communitarian underpinnings, with emphasis on how people ought to behave towards one another as well as the environment. The new world will value altruism, participation and caring. The language was ethically charged, referring to places that 'dare to innovate' and their 'pervasive volunteer spirit'. It offered strong prescriptions as to what attitudes and behaviours were expected in this movement towards the sustainable community.

Naturally, there are similarities between these extracts, which include:

- a sense that the existing city has severe problems which require adoption of new principles of planning and management;
- a conviction that the ideas being put forward are correct and necessary to arrest further deterioration of the environment;
- an emphasis on the human scale. Putting people in touch with their environment means resurrecting the local scale in human affairs as well as recognising global responsibilities;
- a sense that environmental awareness is a basic ingredient of good citizenship. There is an interlinking of arguments about the natural environment and the social environment in which people live. It assumed that if people enjoy a better and more caring relationship with nature, they would also enjoy a better, more caring, and potentially more just relationship with one another;
- a conviction that sustainability when applied to cities is interwoven with diversity and community;
- the direct suggestion that the way forward depends on possessing the appropriate vision. 'Vision' here implies not only being able to visualise what is required but also a strong moral sense that it is correct.

Answering whether or not these extracts are 'utopian' immediately confronts the pejorative uses of the term. Environmentalists may well object in principle to any such description of their cherished ideas because they take it to mean 'unrealistic'. They may interpret matters as meaning that, at the current state-of-the-art, these extracts supply sets of propositions about the world rather than blueprints for replication. Despite these provisos, there are certainly sufficient features in all these extracts that correspond to the description of utopianism provided earlier:

- Each, by the nature of its preferred future, reveals a profound dissatisfaction with the existing city.
- Each is premised on a comprehensive vision that links together people, role, resources and ways-of-life. There is no apparent place for dissent or alternative ways of constructing the future.
- Each extract defines the Good Life by reference to an improved relationship between people and their natural environment. Somehow making best use of resources in the long-term becomes attached to achieving more harmonious societies.

- Each is about 'cities', but statements about future urban environments are secondary to statements about future social environments. Once again, utopianism primarily emphasises social transformation rather than transformation in the physical environment of the city.

Following on from this:

- Each contains propositions about the conditions necessary to achieve the Good Life, without providing a road map of how these ideal conditions are to be achieved. Present-day urban problems, including poverty, discrimination and social deprivation, are noticeable by their absence.
- Each can be employed normatively. They provide standards that can show the difference between what *is* and what *could be*. As such, they provide the basis for critical commentary on the present and its perceived deficiencies.
- Each fails to spell out the political choices necessary to achieve the desired ends. The third extract, for example, adds 'economic self-reliance' to the list of desirable qualities. This, in itself, is a contentious phrase premised on an unspecified standpoint on the relationship between economic, social and environmental well-being.

Taken together, these extracts point to a style of thought and modes of representation that are freshly constructed but familiar. The sustainable city is an alluring place. As Herbert Girardet (1999, 73) summarised:

> If we get things right, cities in the new millennium will be centres for a culture of sustainability. They will be energy- and resource-efficient, people-friendly, and culturally rich, with active democracies assuring the best use of human energies. Prudent infrastructure development will enhance employment, improving public health and living conditions.

The emphasis, however, must be placed on the idea of 'getting things right'. The picture is created of a society living at peace with itself and with nature in resource-efficient surroundings, with a vibrant, inclusive and inspiring public life. The sustainable city is clearly a crucible for social transformation, yet the processes involved in that transformation are not spelled out. The reliance again is on vision. Girardet continued:

> None of this will happen unless we create a new balance between the material and the spiritual, and to that effect much good work needs to be done. A calmer, serener vision of cities is needed to help them fulfil their true potential as places not just of the body but of the spirit.

These are ideas with which Plato would have concurred, but one is still confronted with the elements missing from such formulations. The problems of the existing city, indeed the obstructions caused by the accretion of many years' worth of non-sustainable architecture and traffic practices, are not part of the equation, presumably removed by the alchemy of change carried out in the right spirit. The constant references to the local scale, for example, suggest that the spirit of neighbourhood is still important (see Chapter 7), but there is little about the layout, structure or functioning of the new neighbourhoods in the sustainable city. With the emphasis as ever in utopian schemes on social transformation, the details about the necessary urban environmental transformation are sketchy. If visions of sustainable cities are to prove more reliable in these matters than utopian predecessors, environmentalists will need to devise further, but equally alluring representations of the future city that show precisely, in words or visual images, what the transitional and final stages might look like.

Conclusion

This chapter, the second of two about urban representations, has focused on historic and future cities. When looking at the historic city, we began by noting how cities may be read as text, emphasising the importance of being able to read the signs and to incorporate them into a convincing explanatory narrative. The first part used a case study of Bath to show the changing ways in which powerful groups represented their power by moulding the physical fabric of the city. This led on to a related discussion of the manner in which historic cities are interpreted by the heritage industry. The case study of the dwellings at Matera showed the way that even the most appalling living conditions could, within living memory, be transformed into themed heritage. We concluded that heritage interpretation, and the multiple meanings that it perforce embraces, has an importance that goes beyond academic curiosity.

The latter half of this chapter switched the focus of attention from representations of historic cities to representations of the future city. To do so, we again built on the duality long found in representations of the city in Western society by introducing the concept of 'utopianism' and its antithesis, 'dystopianism'. The discussion then centred on science-fiction film, as a prime example of providing dystopic representations of the future city. Science-fiction film was premised on the principle of action being set against the dystopian background of oppressive, polluted and overcrowded cities of the future. By contrast, writers on sustainable cities embraced the characteristic strategies of utopian writing when representing their ideas about the future city and its cherished characteristics – diversity, community and environmental responsibility.

Further reading

On landscapes of power, see:

David Harvey (2001) *Spaces of Capital: towards a critical geography*, Edinburgh: Edinburgh University Press.
Andy Merrifield (2002) *Dialectical Capitalism: social struggles in the capitalist city*, New York: Monthly Review Press.
Don Mitchell (2003) *The Right to the City: social justice and the fight for public space*, New York: Guilford Press.

On urban heritage and interpretation, see:

Peter Howard (2003) *Heritage: management, interpretation, identity*, London: Continuum.
David Lowenthal (1985) *The Past Is a Foreign Country*, Cambridge: Cambridge University Press.

On utopianism, dystopianism and the city, see:

Peter Hall (2002) *Cities of Tomorrow: an intellectual history of urban planning and design in the twentieth century*, 3rd edn, Oxford: Blackwell.
David Harvey (2000) *Spaces of Hope*, Edinburgh: Edinburgh University Press, 133–96.

On science-fiction cities, see:

Annette Kuhn, ed. (1999) *Alien Zone II: the spaces of science-fiction cinema*, London: Verso.
David Neumann, ed. (1999b) *Film Architecture: set designs from Metropolis to Blade Runner*, Munich: Prestel.

On sustainable cities, see:

Barbara Eckstein and James A. Throgmorton, eds (2003) *Story and Sustainability: planning, practice and possibility for American cities*, Cambridge, MA: MIT Press.
Peter Hall and Colin Ward (1998) *Sociable Cities*, Chichester; John Wiley.
Antonia Layard, Simin Davoudi and Susan Batty, eds (2001) *Planning for a Sustainable Future*, London: E. and F.N. Spon.

9 Conclusion

In this book, we have explored varying ways in which people have represented the environment around them and shown how their representations relate to issues involving environmental attitudes, values and actions. In the process, we have argued that such representations:

- take a wide range of forms, including designed landscapes, scientific and academic writings, newspaper reports, maps and diagrams in addition to paintings, novels, poetry, film and television;
- should be studied for their form, style and structure as well as content;
- convey most information about environmental attitudes and values if they are studied in their historical and social contexts;
- have meanings that may change over time and be disputed by different individuals and interest groups;
- involve practical activity and, as such, representation is itself action and not just the reflection of actions;
- always involve processes of production, transmission and consumption, and include groups of producers, audiences and their institutional contexts;
- have their form and meanings made and shaped by the actions of both producers and consumers;
- are more than insubstantial ideas, but are themselves tangible manifestations of culture and its values.

It has been argued that representations are important for the study of society–environment relationships because:

- they are useful sources for the study of the attitudes and values that inform environmental actions, for example, the use and misuse of the land and the earth's resources, the making and remaking of urban environments;
- environmental representations in the form of, say, landscape design, agricultural practices, scientific investigations, urban planning and environmental protests, are both symbolic and practical interventions in the environment;

- representations of nature, landscape, distant and utopian environments are frequently used as a source of authority to justify a wide range of social, moral and political programmes;
- representations of the city in literature, film, posters, television and other media allow us to scrutinise key discourses that have shaped the everyday urban environment.

It is only too easy to think of environmental representations as passive depictions of the 'real' world and merely partial reflections of reality, because they are filtered and distorted by human interests, aspirations, and incomplete knowledge. This book has repeatedly stressed that this is inadequate when thinking about representation. The process of producing representations, making and interpreting social meanings is an active constituent in the making of environments. Paintings of the Alps or the Lake District, no less than maps and designs for gardens and suburbs, actively shape our reactions to places. Though commonsense suggests that the physical world and the imaginative world of ideas have separate existences, this book has argued that basic categories for understanding the physical world, such as the category 'nature', are in fact cultural constructs. There is therefore a great deal more at stake than simply images and words when making and using environmental representations.

Many modes of representation run deep in Western history. Among those that have recurred at various points in this book, we might include:

1   *Vision as an act of possession*. Chapters 4–5 provided evidence for the power of the visual when showing how landscape, as an elite way of seeing, dictated ideas of nature and controlled their use against the rival claims of popular conceptions of landscape and nature rooted in everyday practical experience. Vision remains central to modern conceptions of knowledge and control. To take some examples:

    - The scientific control of nature is linked to vision, for instance, through measurement in geometry and expressed in graphic and cartographic representation.
    - The objectification of nature in visual form remains fundamental to our ideas of land ownership and property.
    - The objectification and possession of exotic landscapes, animals and plants in photography is important for conservation strategies, tourist economies and the entertainment industries.

2   *Stereotyping and the representation of 'Others'*. As shown by the imperialist representation of nature in Chapter 6 or the representation of urban neighbourhoods in Chapter 7, modes of environmental representations are allied to control. Interpreting the unfamiliar in familiar terms and negatively

contrasting people and places with ourselves creates stereotypes that justify the exercise of power. By these means, the unfamiliar can be subjected and controlled as somehow less than us. To that extent, stereotyping:

- justifies the subjection of nature, women, non-Western peoples and the socially disadvantaged;
- helps perpetuate systematic inequalities;
- transforms social values into a scientific form and thereby renders them legitimate.

3   *Utopian spaces*. Visions of idealised environments come in many shapes and sizes – eighteenth-century landscape gardens, romantic images of mountain communities, isolated exotic islands, and cities ruled by philosophy, rational planning or by environmental responsibility. These have all supplied the territory for playing out critical and alternative ideas of human–environment relationships. The idea of utopia is both an important historical and distinctly modern form of environmental imagining, whether as a blueprint for the future or as supplying normative models that allow us to learn about the deficiencies of the present.

Representation and hybridity

In drawing this book to a close, we note again that environmental representations challenge the conventional separation of image and reality while, at the same time, posing difficult moral and political questions concerning the process of representation and its wider implications. As Chapter 3 indicated, recent debate about the relationships between science, technology, society and environment has pointed to the significance of *hybridity* when searching for ways to face up to the simultaneously imaginative and physical quality of representations. We used the concept of *landscape* as a set of *discursive practices* that highlight the *hybrid* nature of society–environment relations as a means of overcoming the still deeply engrained conceptual separation of nature and culture. This supplies the most accessible route through difficult conceptual material, but this area of debate continues to evolve rapidly. There are, of course, many other topics and perspectives worth studying in pursuing further issues raised in this book. A good example concerns the relationship between representation and hybridity.

Current global environmental problems, like ozone depletion, global warming and deforestation, challenge the separation of nature and society. It is impossible to address such problems by interpreting them simply as either natural or social. Rather it is important to recognise that society and nature are irrevocably fused together by these problems. If we are to cope with the contemporary world, we must embrace hybridity. Writers such as Bruno Latour (e.g. 1993, 1999) and others (e.g. Law, 1991; Law and Hassard, 1999) believe that this requires a profound conceptual reappraisal of the relationship between society and nature. This, in turn, has implications for the history of environmental relations and the conduct of environmental research, action, management and representation. Representational strategies that begin to accomplish this task can be considered under two headings.

Symbolic reconnection

Accusations about the 'culturally impoverished' quality of modern life are a familiar part of many environmentalist critiques of urbanised industrialism (see Milton, 1996). Many environmentalists, especially those concerned with linking community and environmental development (see Chapter 8), emphasise the cultural and symbolic enrichment of environmental experience (Gablik, 1991). The techniques of symbolic reconnection relate closely to the conflict between elite and folk-based conceptions of landscape discussed in Chapters 5 and 6. Frequently associated with reasserting folk-based conceptions of landscape, symbolic reconnection concerns reclaiming lost or suppressed histories in the face of dominant forms of environmental representation.

In the United Kingdom, for example, the organisation Common Ground urges using landscape to reconnect our environmental awareness with forgotten history and neglected cultural meanings (Mabey, 1980). The idea of landscape as a 'theatre of memory' – the hybrid material expression of history, collective identity and social values – has become an increasingly important focus for artists and environ-mentalists. Common Ground's Parish Maps project, for instance, asked local groups to produce symbolic representations of their locality using techniques, materials and forms with which they felt 'at home'. This, they argued, enabled groups to take possession of the ways in which they are represented, redefining their community in their own terms while also mobilising, assessing and galvanising the social and physical resources at the community's disposal (Crouch and Matless, 1996). Artists working alone also draw on this strategy. Tim Robinson's maps and books on the Aran Isles and Burren region in the west of Ireland told the history of human engagement with land and landscape through the history and etymology of place names (Robinson, 1986, 1995). The result recovered a rich tapestry of environmental actions and experiences reaching back over many generations. Work by the sculptor Shelley Sacks on the banana trade put producers and consumers, separated by huge physical and economic distances, in touch with each other in a bid to establish mutual understanding while simultaneously exposing the unequal terms of international trade (Cook, 2000).

Using performance as representation, the London-based radical arts collective 'Platform' employed community-based creative and consciousness raising strategies to highlight environmental and political inequalities. Their *Still Waters* project, for instance, involved a multidisciplinary group of specialists as well as members of the community as observers and participants. The project drew attention to the lost history of London's river valleys in order to increase awareness of the loss of identity and community control caused by generations of urban development. One part of the project, 'Unearthing the Effra', engaged a performance artist and publicist to develop a campaign to dig up the River Effra, buried under suburban south London and restore its natural forms. Another part of the project adapted an old mill to generate hydroelectricity for a primary school (Kastner and Wallis, 1998, 39–40). Environmental interventions by protest groups such as 'Reclaim the Streets' sometimes involve 'adopting' road space by 'planting' trees, making gardens and children's play areas. Such active re-appropriations of road space consciously oppose the current conventions of urban life. Their actions materially and symbolically reconnect with past forms of social life and environmental experience in order to show how government policy and contemporary capitalism have impoverished the lives of city dwellers (Seel et al., 2000; Jordan, 2002). By these means, environmentalists, artists and activists use symbolic identification to encourage practical action.

Representing an active nature

The concept of hybridity also raises issues about human dominance over and control of nature. Rather than solely focusing on humans actively moulding and modifying passive nature, hybridity turns attention to the physical world's involvement in social and cultural systems. Such thinking partly returns the discussion to the concepts of animate nature discussed in Chapter 4. As part of a more general questioning of modern values and practices, these moves are shadowed, on the one hand, by 'new age' forms of environmental spirituality and, on the other, by the mystical systems theory of Gaian environmental philosophy and science. Drawing on an active conception of nature, several authors have provided environmental histories describing the relationship between culture and key natural resources. Henry Hobhouse's study *Seeds of Change* (1985), for example, was of 'six plants which transformed' humankind. It traced the 'causal impact' on human societies of quinine, sugar, tea, cotton, the potato and coca. Likewise, Kurlansky's book *Salt* (2003) attempted to show how a mineral necessary for the preservation and flavouring of food has 'shaped civilisation'. He argued that salt has caused wars, financed empires across Europe and Asia and inspired revolution. From this standpoint, Gandhi's 1930 salt march began the overthrow of British rule in India.

Substantially such work centres on the unintended consequences that lie at the intersection of natural and social systems. Its salutary effect is to show limits to human control of the natural world and to resituate humans back within nature. However provocative the conclusions, these authors criticise, at least implicitly, a mode of thinking that directly stems from the scientific revolution. Artists too have increasingly concerned themselves with the mutually determining effects of human and natural action. In environmental art, ideas of active nature are a means of both highlighting the 'inherent beauty' of nature and recognising the limits of individual human creativity. Since the late 1970s artists such as David Nash have produced objects in wood that crack and twist as the wood dries, seasons and ages or, like Andy Goldsworthy, have made items from fragile materials such as snow, mud, leaves and sand that assume a life of their own as they decay through time (Revill, 1993; Matless and Revill, 1995).

Some have gone further, positing a rethinking of society–nature relationships that attribute some form of will or agency to plant and animal species and communities. This is a highly contentious issue given that it is logically and physically impossible for lower order life forms to exhibit conscious free will. Pollan's *The Botany of Desire* (2002, xii), for example, viewed the domestication of plants in a similar manner to animals, seeing the relationship between society and plants as one of reciprocity or 'co-evolution'. Pollan argued that although we automatically think of domestication as something we do to other species, it makes as much sense to think of it as something certain plants and animals have done to us, a clever evolutionary strategy for advancing their own interests. 'The species that have spent the last ten thousand or so years figuring out how best to feed, heal, clothe, intoxicate, and otherwise delight us have made themselves some of nature's greatest success stories (Pollan, 2002, xiv). By attributing the active will of these species to the effects of evolution, Pollen cleverly avoided crude anthropomorphising of plant and animal life.

The active agency of natural systems and material objects has been a central theoretical issue for those associated with actor-network-theory, which provides a rather more theoretically elaborate approach to representing an active nature. Actor-network-theory is concerned with the way in which, for example, people, artefacts and texts join together to form hybrid structures or networks of socially meaningful activity. The idea of a network in actor-network-theory parallels the concept of a discourse as bundle of symbolic and practical actions, activities, objects and texts (see Chapter 3). Michel Callon (1986), for instance, described a scientific and economic controversy concerning the causes for the decline in the population of scallops in St Brieuc Bay (France) and the attempts of three marine biologists to develop a conservation strategy for that population. He developed a methodology based on the process of 'translation' in order to understand and analyse the attempts by a group of researchers to impose themselves and their definition of the situation on others. In actor-network-theory, translation is the process by which objects, people and representations are 'moved' (translated) from one network (or discourse) to another, or interpreted and claimed as belonging by one network or another by those forces trying to construct a particular network. The novelty of Callon's approach was that he treated plants, animals, insensate objects, books, earth and atmosphere with the same mode of analysis applied to humans. Somewhat playfully, he examined the parallel 'political' constituencies of fishermen and scallops that the scientists attempted to engage, 'persuade' and enlist into their management plans. At the end of Callon's story – and despite the very best attempts of scientists, fishermen and politicians – the scallops migrated down the coast leaving St Brieuc Bay scallop-less. Regardless of how one reacts to this exposition, Callon provided a significant warning that human control of the physical world is always and everywhere limited, even when all available scientific and technical resources are marshalled to exert human authority. From this perspective, an active nature is intimately related to the limitations of representation. In other words, however one tries to describe, classify, measure and theorise animate and inanimate objects, the subjects of our attention always exceed the capacity of any symbolic system to represent them. This is a humbling thought but an important one for the study of environmental representations, because it emphasises that the world is always bigger than our ability to describe, control and predict how it will behave.

The feminist environmentalist Donna Haraway (1990) has shown, for example, that environmental metaphors do not just invoke the negative stereotypes of the type discussed in Chapter 6, since metaphors can also be used strategically and to political effect. Celebrating the medical technologies of birth, birth control and genetic manipulation that render women's lives increasingly hybrid, she developed an image of the cyborg. The cyborg was 'a cybernetic organism, a hybrid of machine and organism, a creature of social reality as well as a creature of fiction' that served as an organising metaphor for identity and environmental politics (Haraway, 1990, 149). Thus contrary to the views of many environmentalists and feminists, Haraway argued that women should use their complex identities in the modern world to break free of their association with the primitive and the natural – itself a legacy of the Baconian conception of science discussed in Chapter 4.

The above examples show the centrality of the representational process to contemporary environmental debate, but introduce representational strategies that are challenging and controversial (e.g. Harvey, 1996; Latour, 1999; Castree and Braun, 2001). Certainly new ways of representing the environment centred on the concept of hybridity provide both opportunities and dangers. On the one hand, it is a deliberate move to take nature seriously and recognise the mutual interdependence of society and nature. Representational strategies that do so can play a positive role in securing environmental quality and diversity. On the other hand, it is easy for humans to persuade themselves, as they have in the past, that they can speak on behalf of nature by devising new means of symbolic representation. Even when we are trying to be as inclusive and even-handed as possible, we must continue to ask 'who has the ability and the right to speak?' and 'whose interests are being represented?' Most of all, there is a need for all those concerned with using and making environmental representations to think through the wider implications of the images and symbols that they choose.

Bibliography

Abercrombie, P. (1939) 'Introduction', in P. Abercrombie, ed. *The Book of the Modern House: a panoramic survey of contemporary domestic design*, London: Waverley, vii–xx.

Ackerman, J.S. (1990) *The Villa: form and ideology of country houses*, London: Thames and Hudson.

Albrecht, D. (1999) 'New York, Olde Yorke: the rise and fall of a celluloid city', in D. Neumann, ed. *Film Architecture: set designs from Metropolis to Blade Runner*, Munich: Prestel, 39–43.

Amin, A. and Thrift, N. (2002) *Cities: reimagining the urban*, Cambridge: Polity Press.

Anderson, A. (1997) *Media, Culture and the Environment*, London: UCL Press.

Anderson, M.D. (1971) *History and Imagery in British Churches*, London: John Murray.

Andrews, M. (1989) *The Search for the Picturesque*, Aldershot: Scolar Press.

Andrews, M. (1999) *Landscape and Western Art*, Oxford: Oxford University Press.

Arnold, D. (1996) *The Problem of Nature: environment, culture and European expansion*, Oxford: Blackwell.

Amin, A. and Thrift, N. (2002) *Cities: reimagining the urban*, Cambridge: Polity Press.

Ashmolean Museum (1988) *Paolo Uccello's 'The Hunt in the Forest'*, Oxford: Ashmolean Museum.

Atkins, P., Simmons, I. and Roberts, B. (1998) *People, Land and Time: an introduction to the relations between landscape, culture and environment*, London: Arnold.

Attfield, R. (1983) 'Christian attitudes to nature', *Journal of the History of Ideas*, 44, 369–86.

Baird Callicot, J. (1994) *Earth's Insights: a multicultural survey of ecological ethics from the Mediterranean basin to the Australian outback*, Berkeley: University of California Press.

Barat, I. (2001) 'Four decades ago', in M.J. Boyd and M.J. Liebler, eds *Abandon Automobile: Detroit city poetry 2001*, Detroit: Wayne State University Press, 55–6.

Barker, C. (2000) *Cultural Studies: theory and practice*, London: Sage.

Barrell, J. (1980) *The Dark Side of the Landscape: the rural poor in English painting, 1730–1840*, Cambridge: Cambridge University Press.

Barthes, R. (1967) *The Elements of Semiology*, London: Cape.

Barthes, R. (1972) *Mythologies*, trans. A. Lavers, London: Paladin.

Bate, J. (1991) *Romantic Ecology: Wordsworth and the environmental tradition*, London: Routledge.

Batey, M. (1982a) 'Landscape Gardens in Oxfordshire', in S. Raphael, ed. *Of Oxfordshire Gardens*, Oxford: Oxford University Press.

Batey, M. (1982b) *Oxford Gardens: the University's influence on garden history*, Amersham: Avebury.

Baudrillard, J. (1994) *Simulacra and Simulation*, trans. S.F. Glaser, Ann Arbor: University of Michigan Press.

Beder, S. (1997) *Global Spin: the corporate assault on environmentalism*, Totnes, Devon: Green Books.

Bender, B. and Winer, M. (2001) *Contested Landscapes: movement, exile and place*, Oxford: Berg.

Berger, J. (1972) *Ways of Seeing*, London: British Broadcasting Corporation and Penguin.

Black, J. (1992) *The British Abroad: the Grand Tour in the eighteenth century*, Cambridge: Cambridge University Press.

Blakely, E.J. and Snyder, M.G. (1997) *Fortress America: gated communities in the United States*, Washington, DC: Brookings Institution Press.

Bocock, R. (1986) *Hegemony*, London: Tavistock Publications.

Borsay, P. (2000) *The Image of Georgian Bath, 1700–2000: towns, heritage, and history*, Oxford: Oxford University Press.

Boulding, K.E. (1966) 'The economics of the coming Spaceship Earth', in H. Jarrett, ed. *Environmental Quality in a Growing Economy*, Baltimore, MD: Johns Hopkins University Press.

Bousé, D. (2000) *Wildlife Films*, Philadelphia: University of Pennsylvania Press.

Branston, G. and Stafford, R. (1996) *The Media Student's Book*, London: Routledge.

Brown, L.A. (1977) *The Story of Maps*, New York: Dover Publications.

Brunn, S.D., Andersson, H. and Dahlman, C.T. (2000) 'Landscaping for power and defence', in J.R. Gold and G. Revill, eds *Landscapes of Defence*, Harlow: Prentice Hall, 68–84.

Bryman, A. (2001) *Social Research Methods*, Oxford: Oxford University Press.

Buckland, A. (2002) 'Woollongong: city of innovation', http://www.wollongong.nsw.gov.au (accessed 30 August 2003).

Bukowczyk, J.J. (1989) 'Detroit: the birth, death and renaissance of an industrial city', in J.J. Bukowczyk and D. Aikenhead, eds *Detroit Images: photographs of the Renaissance City*, Detroit: Wayne State University Press, 15–26.

Bullock, A.L.C. (1988) 'Hegemony', in A. Bullock, ed. *The Fontana Dictionary of Modern Thought*, 2nd edn, London: Fontana, 379.

Bunn, D. (1994) '"Our wattled cot": mercantile and domestic space in Thomas Pringle's African Landscapes', in W.J.T. Mitchell, ed. *Landscape and Power*, Chicago: University of Chicago Press, 127–74.

Burke, P. (2001) *Eyewitnessing: the uses of images as historical evidence*, London: Reaktion Press.

Callon, M. (1986) 'Some elements of a sociology of translation: domestication of the scallops and the fishermen of St Brieuc Bay', in J. Law, ed. *Power, Action and Belief: a new sociology of knowledge?*, London: Routledge and Kegan Paul.

Cameron, J.M.R. (1974) 'Information distortion in colonial promotion: the case of the Swan River colony', *Australian Geographical Studies*, 12, 57–76.

Campbell, C. (1717–25) *Vitruvius Britannicus, or the British Architect, containing the plans, elevations, and sections of the regular buildings, both publick and private, in Great Britain, with variety of new designs*, 3 vols, London: Colin Campbell.

Carey, J. (1992) *The Intellectuals and the Masses: pride and prejudice among the literary intelligentsia, 1880–1939*, London: Faber and Faber.

Carson, R. (1963) *Silent Spring*, Harmondsworth: Penguin.

Carter, Paul (1987) *The Road to Botany Bay*, London: Faber.

Castree, N. and Braun, B. (2001) *Social Nature: theory, practice and politics*, Oxford: Blackwell.

Ceram, C.W. (1965) *Archaeology of the Cinema*, trans. R. Winston, New York: Harcourt, Brace and World.

Chapman, R.W., ed. (1924) *Johnson's 'Journey to the Western Islands of Scotland' and Boswell's 'Journey of a Tour to the Hebrides with Samuel Johnson, LL.D.'*, London: Oxford University Press.

Cherry, G.E. (1988) *Cities and Plans: the shaping of urban Britain in the nineteenth and twentieth centuries*, London: Arnold.

Clark, K. (1976) *Landscape into Art*, London: John Murray.

Clarke, J.J., ed. (1993) *Voices of the Earth: an anthology of ideas and arguments*, London: Earthscan.

Clayre, A., ed. (1977) *Nature and Industrialization*, Oxford: Oxford University Press.

Coates, P. (1998) *Nature: Western attitudes since ancient times*, Cambridge: Polity Press.

Cobbett, W. (1821) *Rural Rides in the Counties of Surrey, Kent, Sussex . . . with economical and political observations, relative to matters applicable to, and illustrated by, the state of the counties, respectively*, London: William Cobbett.

Cohen, I.J. (1989) *Structuration Theory: Anthony Giddens and the constitution of social life*, Basingstoke: Macmillan.

Coleman, B.I., ed. (1973) *The Idea of the City in Nineteenth-Century Britain*, London: Routledge and Kegan Paul

Coleman, J. (1984) 'St Augustine: Christian political thought at the end of the Roman Empire', in British Broadcasting Corporation, *Political Thought from Plato to Nato*, London: Ariel Books/British Broadcasting Corporation, 45–69.

Comune di Matera (2003) 'The Cave Dwelling of Vico Solitario', http://www.casagrotta.it/english.html, (accessed 11 August 2003).

Cook, I. (2000) 'Social sculpture and connective aesthetics: Shelley Sacks's 'exchange values', *Ecumene*, 7, 337–43.

Coones, P. and Patten, J. (1986) *The Penguin Guide to the Landscape of England and Wales,* Harmondsworth: Penguin.

Cosgrove, D. (1984) *Social Formation and the Symbolic Landscape*, London: Croom Helm.

Cosgrove, D. (1993) *The Palladian Landscape*, Leicester: Leicester University Press.

Cosgrove, D. and Daniels, S., eds (1988) *The Iconography of Landscape: essays on the symbolic representation, design and use of past environments*, Cambridge: Cambridge University Press.

Cowper, W. (1785) *The Task, a poem in six books*, London: J. Johnson.

Crabbe, G. (1783) *The Village: a poem. In two books*, London: J. Dodsley.

Crandell, G. (1993) *Nature Pictorialized: 'the view' in landscape history*, Baltimore, MD: Johns Hopkins University Press.

Crang, M. (1994) 'On the heritage trail: maps and journeys to olde Englande', *Environment and Planning D: Society and Space*, 12, 341–55.

Crouch, D. and Matless, D. (1996) 'Refiguring geography: Parish maps of common ground', *Transactions of the Institute of British Geographers*, 21, 236–55.

Culler, J. (1976) *Saussure*, Glasgow: Fontana/Collins.

Culler, J. (1983) *Barthes*, Oxford: Oxford University Press.

Cunliffe, B. (1971) *Roman Bath Discovered*, London: Routledge and Kegan Paul.

Dale, S. (1996) *McLuhan's Children: the Greenpeace message and the media*, Toronto: Between the Lines.

Daley, M.L. (2003) 'Detroit', in S.I. Kutler, ed. *Dictionary of American History*, vol. 3, New York: Charles Scribner's Sons, 19–21.

Daniels, S. (1993) *Fields of Vision: landscape imagery and national identity in England and the United States*, Cambridge: Polity Press.

Darby, W.J. (2000) *Landscape and Identity: geographies of nation and class in England*, Oxford: Berg.

Davis, M. (1992) *City of Quartz: excavating the future in Los Angeles*, New York: Vintage.

Day, J.N. (2003) 'New York City', in S.I. Kutler, ed. *Dictionary of American History*, vol. 6, New York: Charles Scribner's Sons, 78–89.

Defoe, D. (1719) *The Life and Strange Surprising Adventures of Robinson Crusoe, of York, Mariner*, London: W. Taylor.

Defoe, D. (1724–7) *A Tour Through the Whole Island of Great Britain*, London: G. Strahan.

Descola, P. and Palsson, G. (1996) *Nature and Society: anthropological perspectives*, London: Routledge.

Dickens, C. (1836–7) *Sketches by Boz*, London: John Macrone.

Dickens, C. (1848) *Dombey and Son*, London: Chapman and Son.

Dickens, C. (1853) *Bleak House*, London: Bradbury and Evans.

Dickinson, R.E. (1955) *The Population Problem of Southern Italy: an essay in social geography*, Syracuse, NY: Syracuse University Press.

Dolan, B. (2002) *Ladies of the Grand Tour*, London: Flamingo.

Domosh, M. (1996) *Invented Cities: the creation of landscape in nineteenth-century New York and Boston*, New Haven, CT: Yale University Press.

Donkin, R. (1976) 'Changes in the Early Middle Ages', in H.C. Darby, ed. *A New Historical Geography of England before 1600*, Cambridge: Cambridge University Press.

Dorling, D. and Fairbairn, D. (1997) *Mapping: ways of representing the world*, Harlow: Longman.

Dougherty, R.J. (1999) 'Citizen', in A.D. Fitzgerald, ed. *Augustine through the Ages: an encyclopedia*, Grand Rapids, MI: William B. Eerdmans Publishing, 194–6.

Duck, S. (1730) *A Poem (The Thresher's Labour)*, London: n.p.

Dudrah, R.K. (2002) 'Birmingham (UK)', *City*, 6, 335–50.

Duncan, J. (1999) 'Dis-orientation: On the shock of the familiar in a far-away place', in J. Duncan and D. Gregory, eds *Writes of Passage: reading travel writing*, London: Routledge.

Eckstein, B. and Throgmorton, J.A., eds (2003) *Story and Sustainability: planning, practice and possibility for American cities*, Cambridge, MA: MIT Press.

Eco, U. (1986) *Travels in Hyperreality*, trans. W. Weaver, London: Pan in association with Secker and Warburg.

Edlin, H.L. (1969) 'Fifty years of forest parks', *Commonwealth Forestry Review*, 48, 113–26.

Ehrlich, P.R. (1968) *The Population Bomb*, New York: Ballantine Books.

Eisinger, P. (2003) 'Reimagining Detroit', *City and Community*, 2, 85–99.

Eliav-Feldon, M. (1982) *Realistic Utopias: the ideal imaginary societies of the Renaissance, 1516–1630*, London: Oxford University Press.

Ellen, R. and Fukui, K., eds (1996) *Redefining Nature: ecology, culture and domestication*, Oxford: Berg.

Evans, K. (1998) *Copse: the cartoon book of tree protesting*, Biddestone, Wiltshire: Orange Dog Productions.

Farley, R., Holzer, H.J. and Danziger, S. (2000) *Detroit Divided*, New York: Russell Sage Foundation.

Fergusson, A. (1973) *The Sack of Bath: a record and an indictment*, Salisbury, Wiltshire: Compton Russell.

Fermor, S. (1993) *Piero Di Cosimo: fiction, invention and fantasia*, London: Reaktion Books.

Fine, S. (1989) *Violence in the Model City: the Cavanagh administration, race relations, and the Detroit riot of 1967*, Ann Arbor: University of Michigan Press.

Firey, W. (1947) *Land Use in Central Boston*, Cambridge, MA: Harvard University Press.

Fisher, C. (1981) *Custom, Work and Market Capitalism: the Forest of Dean Colliers, 1788–1888*, London: Croom Helm.

Fishman, R. (1977) *Urban Utopias in the Twentieth Century: Ebenezer Howard, Frank Lloyd Wright and Le Corbusier*, New York: Basic Books.

Fishman, R. (1984) *Urban Utopias in the Twentieth Century*, New York: Basic Books.

Fishman, R. (1987) *Bourgeois Utopias: the rise and fall of suburbia*, New York: Basic Books.

Fried, M. (1963) 'Grieving for a lost home', in L.H. Duhl, ed. *The Urban Condition*, New York: Basic Books, 151–71.

Fried, M. and Gleicher, P. (1961) 'Some sources of residential satisfaction in an urban slum', *Journal of the American Institute of Planners*, 27, 305–15.

Fuller, P. (1988) 'The geography of Mother Nature', in D. Cosgrove and S. Daniels, eds *The Iconography of Landscape: essays on the symbolic representation, design and use of past environments*, Cambridge: Cambridge University Press.

Fuller, P. (1989) 'Fine Arts' in B. Ford, ed. *Victorian Britain,* vol. 7, *The Cambridge Cultural History of Britain*, Cambridge: Cambridge University Press.

Gablik, S. (1991) *The Reenchantment of Art*, London: Thames and Hudson.

Gay, J. (1716) *Trivia: or the art of walking the streets of London*, London: Bernard Lintott.

Geertz, C. (1973) *The Interpretation of Cultures: selected essays*, New York: Basic Books.

Giddens, A. (1984) *The Constitution of Society: introduction of the theory of structuration*, Berkeley: University of California Press.

Giles, J. and Middleton, T. (1999) *Studying Culture: a practical introduction*, Oxford: Blackwell.

Girardet, H. (1999) *Creating Sustainable Cities*, Schumacher Briefing 2, Totnes, Devon: Green Books for the Schumacher Society.

Glacken, C. (1967) *Traces on the Rhodian Shore: nature and culture in Western thought from ancient times to the end of the eighteenth century*, Berkeley: University of California Press.

Gogermany (2003) 'Matera, Italy', http://gogermany.about.com/cs/materaitaly (accessed 11 August 2003).

Gold, J.R. (1974) *Communicating Images of the Environment*, Occasional Paper 29, Centre for Urban and Regional Studies, University of Birmingham.

Gold, J.R. (1980) *An Introduction to Behavioural Geography*, Oxford: Oxford University Press.

Gold, J.R. (1994) 'Locating the message: place promotion as image communication', in J.R. Gold and S.V. Ward, eds *Place Promotion: the use of publicity and public relations to sell towns and regions*, Chichester: John Wiley, 19–37.

Gold, J.R. (1997) *The Experience of Modernism: modern architects and the future city, 1928–53*, London, E. and F.N. Spon/Routledge.

Gold, J.R. (2001) 'Under darkened skies: the city in science-fiction film', *Geography*, 86(4), 337–45.

Gold, J.R. and Gold, M.M. (1990) '"A Place of Delightful Prospects": promotional imagery and the selling of suburbia', in L. Zonn, ed. *Place Images in the Media: portrayal, meaning and experience*, Savage, MD: Rowman and Littlefield, 159–82.

Gold, J.R. and Gold, M.M. (1994) '"Home at last!": building societies, home ownership and the rhetoric of suburban place promotion', in J.R. Gold and S.V. Ward, eds *Place Promotion: the use of publicity and public relations to sell towns and regions*, Chichester: John Wiley, 75–92.

Gold, J.R. and Gold, M.M. (1995) *Imagining Scotland: tradition, representation and promotion in Scottish tourism since 1750*, Aldershot: Scolar Press.

Gold, J.R. and Gold, M.M. (2005) *Cities of Culture: staging international festivals and the urban agenda, 1851–2000*, Aldershot: Ashgate Press.

Gold, J.R. and Revill, G.E., eds (2000) *Landscapes of Defence*, Harlow: Prentice Hall.

Gold, J.R. and Ward, S.V., eds (1994) *Place Promotion: the use of publicity and public relations to sell towns and regions*, Chichester: John Wiley.

Gold, M. (1984) 'A history of nature', in D. Massey and J. Allen, eds *Geography Matters! A reader*, Cambridge: Cambridge University Press, 12–33.

Gold, R. (1971) 'Urban violence and contemporary defensive cities', in W. McQuade, ed. *Cities fit to live in*, London: Collier-Macmillan, 4–20.

Goldsmith, O. (1770) *The Deserted Village, a poem*, London: W. Griffin.

Goodey, B. (1994) 'Interpretive planning', in R. Harrison, ed. *Manual of Heritage Management*, London: Butterworth Heinemann.

Goodwin, B. (1978) *Social Science and Utopia: nineteenth century models of social harmony*, Hassocks, Sussex: Harvester Press.

Gottdiener, M. (1995) *Postmodern Semiotics: material culture and the frame of postmodern life*, Oxford: Blackwell.

Gronow, J. (1997) *The Sociology of Taste*, London: Routledge.

Guthrie, J. (2003) 'Birmingham to seek life beyond the Bullring', *Financial Times*, 10 July, 13.

Hadfield, M. (1977) *The English Landscape Garden*, Aylesbury, Buckinghamshire: Shire Publications.

Hall, P. (2002) *Cities of Tomorrow: an intellectual history of urban planning and design in the twentieth century*, 3rd edn, Oxford: Blackwell.

Hall, P. and Ward, C. (1998) *Sociable Cities*, Chichester: John Wiley.

Hall, S., ed. (1997) *Representation: cultural representations and signifying practices*, London: Sage/Open University.

Hall, T. and Hubbard, P., eds (1996) *The Entrepreneurial City: geographies of politics, regime and representation*, Chichester: John Wiley.

Hannibal, J. (2001) 'Triangulation: a guide', http://web.isp.cz/jcrane/IB/triangulation.html

Hannigan, J.A. (1995) *Environmental Sociology: a social constructionist perspective*, London: Routledge.

Haraway, D. (1990) *Simians, Cyborgs and Women: the reinvention of nature*, London: Free Association Books.

Hardin, G. (1968) 'The tragedy of the commons', *Science*, 162 (1243), 88, 89, 92.

Harley, J.B. (1988) 'Maps, knowledge, and power', in D. Cosgrove and S. Daniels, eds *The Iconography of Landscape: essays on the symbolic representation, design and use of past environments*, Cambridge: Cambridge University Press, 277–312.

Harre, R., Brockmeier, J. and Muhlhausler, P. (1999) *Greenspeak: a study of environmental discourse*, London: Sage.

Harrison, R. (1994) 'Telling the story in museums', in J.M. Fladmark, ed. *Cultural Tourism*, Wimbledon: Donmark, 311–21.

Hart, C. (1971) *The Industrial History of Dean,* Newton Abbot, Devon: David and Charles.

Harvey, D. (1996) *Justice, Nature and the Geography of Difference*, Oxford: Blackwell.

Harvey, D. (2000) *Spaces of Hope*, Edinburgh: Edinburgh University Press.

Harvey, D. (2001) *Spaces of Capital: towards a critical geography*, Edinburgh: Edinburgh University Press.

Harvey, J. (1981) *Medieval Gardens*, London: Batsford.

Hauptman, W. (1991) *Magnificent Switzerland: views by foreign artists 1770–1914*, Lugano: Fondazione Thyssen-Bornemisza.

Hawthorne, S., ed. (1870) *Passages from the English Notebooks of Nathaniel Hawthorne*, 2 vols, Boston: Fields, Osgood and Co.

Hemstedt, G. (1996) 'Inventing social identity: *Sketches by Boz*, in R. Robbins and J. Wolfreys, eds *Victorian Identities: social and cultural formations in nineteenth-century literature*, Basingstoke: Macmillan, 215–29.

Hewison, R. (1987) *The Heritage Industry: Britain in a climate of decline*, London: Methuen.

Hibbert, C. (1974) *The Grand Tour*, London: Spring Books.

Hirsch, E. and O'Hanlon, M. (1995) *The Anthropology of Landscape: perspectives on place and space*, Oxford: Clarendon Press.

Hobhouse, H. (1985) *Seeds of Change*, London: Sidgwick and Jackson.

Holcomb, B. (1994) 'City make-overs: marketing the post-industrial city', in J.R. Gold and S.V. Ward, eds *Place Promotion: the use of publicity and public relations to sell towns and regions*, Chichester: John Wiley, 115–32.

Hooke, J.M. and Kain, R.J.P. (1982) *Historical Change in the Physical Environment: a guide to sources and techniques*, London: Butterworth Scientific.

Howard, P. (2003) *Heritage: management, interpretation, identity*, London: Continuum.

Hoy, D.C. (1986) *Foucault: a critical reader*, Oxford: Blackwell.

Hughes, J.D. (1994) *Pan's Travail: environmental problems of the Ancient Greeks and Romans*, Baltimore, MD: Johns Hopkins University Press.

Hughes, R. (1980) *The Shock of the New: art and the century of change*, London: Thames and Hudson.

ISC (Institute for Sustainable Communities) (2003) 'What is a sustainable community?', http://www.iscvt.org/aboutiscgoto.html (accessed 30 August 2003).

Italian Tourism (2003) 'Discover: I Sassi di Matera – Basilicata', http://www.italian tourism.com/discov2.html (accessed 11 August 2003).

Jackson, N. (1991) *Nineteenth-Century Bath Architects and Architecture*, Bath: Ashgrove Press.

Jacobs, J. (1962) *The Death and Life of Great American Cities*, London: Cape.

Jacques, D. (1983) *Georgian Gardens: the reign of nature*, London: Batsford.

Jamison, A. (2001) *The Making of Green Knowledge: environmental politics and cultural transformation*, Cambridge: Cambridge University Press.

Jarvis, R. (1997) *Romantic Writing and Pedestrian Travel*, Basingstoke: Macmillan.

Jellicoe, G. and Jellicoe, S. (1998) *The Landscape of Man: shaping the environment from prehistory to the present day*, London: Thames and Hudson.

Jensen Adams, A. (1994) Seventeenth-century Dutch landscape painting', in W.J.T. Mitchell, ed. *Landscape and Power*, Chicago: University of Chicago Press, 35–76.

Johnston, R.J., Gregory. D., Pratt, G. and Watts, M., eds (2000) *The Dictionary of Human Geography*, Oxford: Blackwell.

Jolly, K. (1993) 'Father God and Mother Earth: nature-mysticism in the Anglo-Saxon world', in J.E. Salisbury, ed. *The Medieval World of Nature*, New York: Garland.

Jones, H.M. (1946) 'The colonial impulse: an analysis of the promotion literature of colonisation', *Proceedings of the American Philosophical Society*, 90, 131–61.

Jordan, G. and Weedon, C. (1995) *Cultural Politics: class, gender, race and the postmodern world*, Oxford: Blackwell.

Jordan, G. and Weedon, C. (2000) *Cultural Politics*, 2nd edn, Oxford: Blackwell.

Jordan, T. (2002) *Activism! Direct Action, Hacktivism and the future of society*, London: Reaktion Books.

Judd, D. (1999) 'Constructing the tourist bubble', in D. Judd and S. Fainstein, eds *The Tourist City*, New Haven, CT: Yale University Press, 35–53.

Kastner, J. and Wallis, B. (1998) *Land and Environmental Art*, London: Phaidon.

Kerner Report (1968) *Report of the National Advisory Commission on Civil Disorders*, Washington DC: US Government Printing Office.

Kingsley, C. (1850) *Alton Locke, Tailor and Poet: an autobiography*, New York: Harper and Brothers.

Kuhn, A., ed. (1999) *Alien Zone II: the spaces of science-fiction cinema*, London: Verso.

Kurlansky, M. (2003) *Salt*, London: Vintage.

Lane, F. (2000) *Pierre Bourdieu: a critical introduction*, London: Pluto.

Latour, B. (1993) *We Have Never Been Modern*, Hemel Hempstead: Harvester Wheatsheaf.

Latour, B. (1999) *Pandora's Hope: essays on the reality of science studies*, Cambridge, MA: Harvard University Press.

Law, J. (1991) *A Sociology of Monsters: essays on power, technology and domination*, London: Routledge.

Law, J. and Hassard, J. (1999) *Actor Network Theory and After*, Oxford: Blackwell.

Lawrence, D.H. (1929) *Nottingham and the Mining Country*, reprinted in D.H. Lawrence (1950) *Selected Essays*, Harmondsworth: Penguin.

Layard, A., Davoudi, S. and Batty, S., eds (2001) *Planning for a Sustainable Future*, London: E. and F.N. Spon.

Layder, D. (1994) *Understanding Social Theory*, London: Sage.

Leake, P. (1912) *History of Detroit: a chronicle of its progress, its industries, its institutions, and the people of the fair City of the Straits,* Chicago and New York: Lewis Publishing Company.

Lees, A. (1985) *Cities Perceived: urban society in European and American thought, 1820–1940*, Manchester: Manchester University Press.

Leonard, C. (1980) *City Primeval: high noon in Detroit*, New York: Arbor House.

Leppert, R. (1996) *Art and the Committed Eye: the cultural functions of imagery*, Boulder, CO: Westview Press.

Levi, P. (1948) *Christ Stopped at Eboli*, London: Cassell and Co.

Ley, D. (1974) *The Black Inner City as Frontier Outpost: images and behaviour of a Philadelphia neighbourhood*, Washington: Association of American Geographers.

Lovelock, J. (1988) *The Ages of Gaia: a biography of our living earth*, New York: Norton.

Lowenthal, D. (1985) *The Past Is a Foreign Country*, Cambridge: Cambridge University Press.

Lowenthal, D. (1996) *The Heritage Crusade and the Spoils of History*, New York: Simon and Schuster.

LSCP (Learning for Sustainable Cities Project) (2003) 'What is a sustainable city', http://www.dep.org.uk/cities (accessed 30 August 2003).

Lynch, K. (1960) *The Image of the City*, Cambridge, MA: MIT Press.

Mabey, R. (1980) *The Common Ground: a place for nature in Britain's future?*, London: Hutchinson.

McArthur, A. and Kingsley Long, H. (1935) *No Mean City*, London: Longmans Green.

McClintock, A. (1995) *Imperial Leather: race, gender and sexuality in the colonial contest*, London: Routledge.

McDowell, L. (1997) *Capital Culture: gender at work in the City*, Oxford: Blackwell.

MacKenzie, J.M. (1988) *The Empire of Nature: hunting, conservation and British imperialism*, Manchester: Manchester University Press.

MacKenzie, J.M., ed. (1990) *Imperialism and the Natural World*, Manchester: Manchester University Press.

McKibbin, R. (1989) *The End of Nature*, New York: Vintage.

Macnaghten, P. and Urry, J. (1998) *Contested Natures*, London: Sage.

McQuail, D. (1997) *Audience Analysis*, London: Sage.

Manuel, F.E. and Manuel, F.P. (1979) *Utopian Thought in the Western World*, Oxford: Blackwell.

Marquis, A.N. (1908) *The Book of Detroiters: a biographical dictionary of leading men of the city of Detroit*, Chicago: A.N. Marquis and Co.

Marx, L. (1964) *The Machine in the Garden: technology and the pastoral idea in America*, Oxford: Oxford University Press.

Matless, D. (1998) *Landscape and Englishness*, London: Reaktion Press.

Matless, D. and Revill, G.E. (1995) 'A solo ecology: the environmental art of Andy Goldsworthy', *Ecumene*, 2, 423–48.

Mazzolani, L.S. (1970) *The Idea of the City in Roman Thought: from walled city to spiritual commonwealth*, London: Hollis and Carter.

Merchant, C. (1995) *Earthcare: women and the environment*, London: Routledge.

Merrick (1996) *Battle for the Trees: three months of responsible ancestry*, Leeds: godhaven ink.

Merrifield, A. (2002) *Dialectical Capitalism: social struggles in the capitalist city,* New York: Monthly Review Press.

Milton, K. (1996) *Environmentalism and Cultural Theory: exploring the role of anthropology in environmental discourse*, London: Routledge.

Mitchell, D. (2003) *The Right to the City: social justice and the fight for public space*, New York: Guilford Press.

Mitchell, W.J.T., ed. (1994) *Landscape and Power*, Chicago: University of Chicago Press.

More, T. (1516) *Utopia*. Version quoted here is E. Surtz and J.H. Hexter, eds (1965) *The Complete Works of Sir Thomas More*, vol. 4, New Haven, CT: Yale University Press.

Morley, D. (1992) *Television, Audiences and Cultural Studies*, London: Routledge.

Mosser, M. and Teyssot, G., eds (1991) *The History of Garden Design: the Western tradition from the Renaissance to the present day*, London: Thames and Hudson.

Muir, R. (1999) *Approaches to Landscape*, Basingstoke: Macmillan.

Mumford, L. (1965) 'Utopia, the city and the machine', *Daedalus*, 94, 270–83.

Mutch, W.E.S. (1968) *Public Recreation in National Forests: a factual survey*, London: HMSO.

Nairn, I. (1961a) 'Gentle, not genteel', *Architectural Review*, 130, 387–93.

Nairn, I. (1961b) 'Spec.-built', *Architectural Review*, 129, 162–70.

Nash, R. (1982) *Wilderness and the American Mind*, New Haven, CT: Yale University Press.

Negley, G.R. and Patrick, J.M. (1952) *The Quest for Utopia: an anthology of imaginary societies*, New York: Schuman.

Neill, W.J.V. (1995) 'Promoting the city: image, reality and racism in Detroit', in W.J.V. Neill, D.S. Fitzsimons and B. Murtagh, eds *Reimaging the Pariah City*, Aldershot: Avebury, 113–61.

Neill, W.J.V., Fitzsimons, D.S. and Murtagh, B., eds (1995) *Reimaging the Pariah City*, Aldershot: Avebury.

Neumann, D. (1999a) 'Before and after Metropolis: film and architecture in search of the modern city', in D. Neumann, (ed) *Film Architecture: set designs from Metropolis to Blade Runner*, Munich: Prestel, 33–8.

Neumann, D., ed. (1999b) *Film Architecture: set designs from Metropolis to Blade Runner*, Munich: Prestel.

Neumann, M. (2001) 'Emigrating to New York in 3-D: stereoscopic vision in IMAX's cinematic city', in M. Shiel and T. Fitzmaurice, eds *Cinema and the City: film and urban societies in a global context*, Oxford: Blackwell, 109–21.

Nicholls, J.C. (2003) 'Gin Lane revisited: intoxication and society in the gin epidemic', *Journal for Cultural Research*, 7, 125–46.

Nicholson, N. (1978) *The Lake District*, Harmondsworth: Penguin.

Nicholson, P. (1984) 'Aristotle: ideals and realities', in British Broadcasting Corporation, *Political Thought from Plato to Nato*, London: Ariel Books/British Broadcasting Corporation, 30–44.

O'Connor, T.H. (1993) *Building a New Boston: politics and urban renewal, 1950–70*, Boston: Northeastern University Press.

Oelschlaeger, M. (1991) *The Idea of Wilderness*, New Haven, CT: Yale University Press.

Olwig, K. (1993) 'Sexual cosmology: nation and landscape at the conceptual interstices of nature and culture; or what does landscape really mean?', in B. Bender, ed. *Landscape: politics and perspectives*, Oxford: Berg, 307–43.

Olwig, K.R. (1996) 'Recovering the substantive nature of landscape', *Annals of the Association of American Geographers*, 86, 630–53.

Olwig, K.R. (2002) *Landscape, Nature, and the Body Politic: from Britain's renaissance to America's new world*, Madison: University of Wisconsin Press.

Osborne, H., ed. (1970) *The Oxford Companion to Art,* Oxford: Oxford University Press.

Ott, F.W. (1979) *The Films of Fritz Lang*, Secanthus, NJ: Citadel Press.

Ousby, I. (1990) *The Englishman's England: taste, travel and the rise of tourism*, Cambridge: Cambridge University Press: Cambridge.

Parker, J. (2000) *Structuration*, Buckingham: Open University Press.

Patterson, A. (1988) *Pastoral and Ideology: Virgil to Valery*, Oxford: Clarendon Press.

Pawson, E. (1992) 'Two New Zealands: Maori and European', in K. Anderson and F. Gale, eds *Inventing Places: studies in cultural geography*, Melbourne: Longman Cheshire, 15–33.

Payne, D.G. (1996) *Voices in the Wilderness: American nature writing and environmental politics*, Hanover, NH: University Press of New England.

Pepper, D. (1996) *Modern Environmentalism: an introduction*, London: Routledge.

Perry, C. (1929) 'The neighbourhood unit', *The Regional Plan for New York and its Environs*, vol. 3, *Neighbourhood and Community Planning*, New York: New York Port Authority.

Pevsner, N. (1958) *The Buildings of England: North Somerset and Bristol*, Harmondsworth: Penguin.

Philo, C. and Kearns, G. (1993) 'Culture, history, capital: a critical introduction to the selling of places', in G. Kearns and C. Philo, eds *Selling Places: the city as cultural capital, past and present*, Oxford: Pergamon Press, 1–32.

Pollan, M. (2002) *The Botany of Desire*, London: Bloomsbury.

Porter, R. (1994) 'Enlightenment London and urbanity', in T.D. Hemming, E. Freeman and D. Meakin, eds *The Secular City: studies in the Enlightenment*, Exeter: University of Exeter Press, 27–41.

Posner, M.L. (2000) 'The Big Apple', in T. Pendergast and S. Pendergast, eds *St James Encyclopedia of Popular Culture*, vol. 1, Farmington Hills, MI: St James Press, 246.

Poster, M. (1998) 'Baudrillard, Jean', in E. Craig, ed. *Routledge Encyclopedia of Philosophy*, vol. 1, London: Routledge, 662–3.

Pradeau, J.F. (1997) *Plato and the City: a new introduction to Plato's political thought*, Exeter: University of Exeter Press.

Powell, R.A. and Single, H.M. (1996) 'Focus groups', *International Journal of Quality in Health Care*, 8, 499–504.

Rackham, O. (2001) *Trees and Woodland in the British Landscape*, London: Phoenix Press.

Reed, M. (1990) *The Landscape of Britain: from the beginnings to 1914*, London: Routledge.

Revill G, (1993) 'The Forest of Dean: art, ecology and the industrial landscape', in J.R. Gold and S.V. Ward, eds *Place Promotion: the use of publicity and public relations to sell towns and regions*, Chichester: John Wiley, 233–45.

Revill, G. and Watkins, C. (1996) 'Educated access: interpreting Forestry Commission Forest Park Guides', in C. Watkins, ed. *Rights of Way: policy, culture and management*, London: Pinter, 100–28.

Robertson, M. (1925) *Laymen and the New Architecture*, London: John Murray.

Robinson, T. (1986) *Stones of Aran: pilgrimage*, Dublin: Lilliput Press.

Robinson, T. (1995) *Stones of Aran: labyrinth*, Dublin: Lilliput Press.

Rodner, W.S. (1997) *J.M.W. Turner: romantic painter of the Industrial Revolution*, Berkeley: University of California Press.

Rogers, P. (1971) 'Introduction' to *A Tour Through the Whole Island of Great Britain*, Harmondsworth: Penguin.

Rogers, R. (1997) *Cities for a Small Planet*, London: Faber and Faber.

Rose, G. (1992) 'Geography as a science of observation: the landscape, the gaze and masculinity', in F. Driver and G. Rose, eds *Nature and Science: essays in the history of geographical knowledge*, London: Institute of British Geographers, Research Series 28, 8–18.

Rose, G. (1993) *Feminism and Geography: the limits of geographical knowledge*, Cambridge: Polity Press.

Rose, G. (2001) *Visual Methodologies*, London: Sage.

Rosenthal, M. (1982) *British Landscape Painting*, Oxford: Phaidon.

Rotella, C. (2003) 'The old neighbourhood', in B. Eckstein and J.A. Throgmorton, eds *Story and Sustainability: planning, practice and possibility for American cities*, Cambridge, MA: MIT Press, 87–110.

Said, E. (1978) *Orientalism*, New York: Pantheon Books.

Sardar, Z. and Loon, B.V. (1997) *Cultural Studies for Beginners*, Cambridge: Icon Books.

Schama, S. (1995) *Landscape and Memory*, London: HarperCollins.

Schwarzbach, F.S. (1979) *Dickens and the City*, London: Athlone Press.

Seel, B., Paterson, M. and Doherty, B., eds (2000) *Direct Action in British Environmentalism,* London: Routledge.

Secondat, C. de, Baron de Montesquieu (1750) *The Spirit of Laws*, London: Nourse and Vaillant.

Selby, K. and Cowdery, R. (1995) *How to Study Television*, Basingstoke: Macmillan.

Sharp, T. (1936) 'The English tradition in the town: III – Universal Suburbia', *Architectural Review*, 79, 115–20.

Shelley, M.W. (1818) *Frankenstein, or the Modern Prometheus*, 3 vols, London: Lackington, Hughes, Harding.

Short, J.R. (1996) *The Urban Order: an introduction to cities, culture and power*, Oxford: Blackwell.

Short, J.R. and Kim, Y.H. (1996) 'Urban crises/urban representations: selling the city in difficult times', in T. Hall, and P. Hubbard, eds *The Entrepreneurial City: geographies of politics, regime and representation*, Chichester: John Wiley, 55–75.

Sinclair, D. (1981) 'Land: Maori view and European response', in M. King, ed. *Te Ao Hurihuri. The World Moves On: aspects of Maoritanga*, Auckland: Longman Paul.

Smith, B. (1985) *European Vision and the South Pacific*, New Haven, CT: Yale University Press.

Smith, H.N. (1970) *Virgin Land: the American West as symbol and myth*, Cambridge, MA: Harvard University Press.

Smith, N. and Godlewska, A., eds (1994) *Geography and Empire*, Oxford: Blackwell.

Soper, K. (1995) *What is Nature?*, Oxford: Blackwell.

Spengler, O. (1926) *The Decline of the West,* 2 vols, trans. C.F. Atkinson, London: George Allen and Unwin.

Stark, G.W. (1943) City of Destiny: the story of Detroit, Detroit: Arnold-Powers, Inc.

Stern, R.A.M., Mellins, T. and Fishman, D. (1995) *New York 1960: architecture and urbanism between the Second World War and the Bicentennial*, New York: Monacelli Press.

Storey, J. (1993) *An Introductory Guide to Cultural Theory and Popular Culture*, Hemel Hempstead: Harvester Wheatsheaf.

Strang, V. (1997) *Uncommon Ground: cultural landscapes and environmental values*, Oxford: Berg.

Suttles, G.D. (1972) *The Social Construction of Communities*, Chicago: University of Chicago Press.

Strong, R. (1984) *Splendour at Court: Renaissance spectacle and illusion*, London: Weidenfeld and Nicolson.

Taylor, C. (1983) *The Archaeology of Gardens*, Aylesbury, Buckinghamshire: Shire Publications.

Terkel, S. (1968) *Division Street, America: report from an American city*, Harmondsworth: Penguin.

Thacker, C. (1994) *The Genius of Gardening: the history of gardens in Britain and Ireland*, London: Weidenfeld and Nicolson.

Thomas, J.M. (1997) *Redevelopment and Race: planning a finer city in postwar Detroit*, Baltimore, MD: Johns Hopkins University Press.

Thomas, K. (1971) *Religion and the Decline of Magic*, London: Weidenfeld and Nicolson.

Thomas, K. (1983) *Man and the Natural World*, Harmondsworth: Penguin.

Thompson, E.P. (1978) 'The poverty of theory or an orrery of errors', in E.P. Thompson, *The Poverty of Theory and Other Essays*, London: Merlin Press, 193–401.

Thompson, E.P. (1997) *The Romantics: England in a revolutionary age*, New York: New Press.

Thompson, F.M.L. (1982) 'Introduction: the rise of suburbia', in F.M.L. Thompson, ed. *The Rise of Suburbia*, Leicester; Leicester University Press, 1–25.

Thompson, J.B. (1984) *Studies in the Theory of Ideology*, Berkeley: University of California Press.

Tod, I.J. (1982) 'The city in Utopia: how might we live?', in L. Taylor, ed. *Cities: the forces that shape them*, New York: Rizzoli and the Cooper-Hewitt Museum, 8–9.

Tuan, Y.-F. (1974) *Topophilia: a study of environmental perception, attitudes, and values*, Englewood Cliffs, NJ: Prentice Hall.

Tuan, Y.-F. (1993) *Passing Strange and Wonderful: aesthetics, nature, and culture*, Washington, DC: Island Press/Shearwater Books.

Tucker, J. (1783) *Four Letters on Important National Subjects, Addressed to the Right Honourable, the Earl of Selbourne*, Dublin: W. and W. Whitestone.

Turner, T.H.D. (1986) *English Garden Design: history and styles since 1650*, Woodbridge, Suffolk: Antique Collectors' Club.

Udall, S.L. (1963) *The Quiet Crisis and the Next Generation*, Salt Lake City, UT: Gibbs and Smith.

UNESCO (2003) 'I Sassi di Matera', http://whc.unesco.org/sites/670.htm (accessed 30 August 2003).

Urry, J. (1990) *The Tourist Gaze: leisure and travel in contemporary societies*, London: Sage.

Urry, J. (1995) *Consuming Places*, London: Routledge.

Van Leeuwen, T. and Jewitt, C., eds (2001) *Handbook of Visual Analysis*, London: Sage.

Vasaly, A. (1993) *Representations: images of the world in Ciceronian oratory*, Berkeley: University of California Press.

Virgil (1983) *The Eclogues: The Georgics*, trans. C.D. Lewis, Oxford: Oxford University Press.

Wallace, A. (1993) *Walking, Literature, and English Culture: the origins and uses of peripatetic in the nineteenth century*, Oxford: Clarendon Press.

Ward, S.V. (1998) *Selling Places*, London: E. and F.N. Spon.

Warde, A. (1994) 'Consumption, identity-formation and uncertainty', *Sociology*, 28, 877–98.

Warner, M. (1998) *No Go the Bogeyman: scaring, lulling, and making mock*, London: Chatto and Windus.

Warnke, M. (1994) *Political Landscape: the art history of nature*, London: Reaktion Books.

Watt, W.C. (1998) 'Semiotics', in E. Craig, ed. *Routledge Encyclopedia of Philosophy*, vol. 8, London: Routledge, 675–9.

WCED (World Commission on Economic Development) (1987) *Our Common Future* (The Brundtland Report), Oxford: Oxford University Press.

Wekerle, G.R. and Whitzman, C. (1995) *Safe Cities: guidelines for planning, design and management*, New York: Van Nostrand Reinhold.

Welsh, A. (1971) *The City of Dickens*, Oxford: Clarendon Press.

Whatmore, S. (2002) *Hybrid Geographies: natures, cultures, spaces*, London: Sage.

Whistler, C. and Bomford, D. (1999) *The Forest Fire*, Oxford: Ashmolean Museum.

White, E.B. (1949) *Here is New York*, New York: Warner Books.

White, L. Jnr (1967) 'The historical roots of our ecological crisis', *Science,* 155, 1203–7.

Whitfield, P. (1994) *The Image of the World: 20 centuries of world maps*, London: British Library.

Whitman, W. (1975) *The Complete Poems*, ed. F. Murphy, Harmondsworth: Penguin.

Whyte, I.D. (2002) *Landscape and History since 1500*, London: Reaktion Books.

Williams, R. (1973) *The Country and the City*, London: Chatto and Windus.

Williams, R. (1976) *Keywords: a vocabulary of culture and society*, London: Fontana.

Williams, R. (1981) *Culture*, London: Fontana.

Williamson, T. (1995) *Polite Landscapes: gardens and society in eighteenth-century England*, Stroud, Glos.: Alan Sutton.

Williamson, T. (2002) *The Transformation of Rural England: farming and the landscape 1700–1870*, Exeter: University of Exeter Press.

Wilson, A. (1992) *The Culture of Nature: North American landscape from Disney to the Exxon Valdez*, Oxford: Blackwell.

Wolfreys, J. (1996) 'Dickensian architextures or, the city and the ineffable', in R. Robbins and J. Wolfreys, eds *Victorian Identities: social and cultural formations in nineteenth-century literature*, Basingstoke: Macmillan, 199–214.

Womersley, D., ed. (1998) *A Philosophical Enquiry into the Origin of our Ideas of the Sublime and Beautiful and other Pre-Revolutionary Writings by Edmund Burke*, Harmondsworth: Penguin.

Wood, D. (1993) *The Power of Maps*, London: Routledge.

Woods, M. (1996) *Visions of Arcadia: European gardens from Renaissance to Rococo*, London: Aurum Press.

Worster, D. (1977) *Nature's Economy: a history of ecological ideas*, Cambridge: Cambridge University Press.

WSU (Washington State University) (2003) 'What is Discourse?', http://www.wsulibs.wsu.edu/electric/trainingmods/academic_disciplines/discourse.htm (accessed 31 August 2003).

Young, E. (1992) 'Hunter-gatherer concepts of land and its ownership in remote Australia and North America', in K. Anderson and F. Gale, eds *Inventing Places: studies in cultural geography*, Melbourne: Longman Cheshire, 255–72.

Films

Across the Sea of Time, dir. Stephen Low (Sony, 1995).

Aelita, Queen of Mars, dir. Jakov A. Protazanov (Mezhrabprom-Rus, 1924).

Algol, dir. Hans K. Werckmeister (Deutsche Lichtspielgesellschaft, 1920).

Alphaville, dir. Jean-Luc Godard, (Chaumiane-Filmstudio, 1965).

Blade Runner, dir. Ridley Scott (Ladd Company/Warner Brothers, 1982).

A Clockwork Orange, dir. Stanley Kubrick (Polaris Productions/Warner Brothers, 1971).

Escape from New York, dir. John Carpenter (Goldcrest, 1981).

Fahrenheit 451, dir. François Truffaut (Anglo-Enterprise Film Productions/Vineyard Productions, 1966).

Logan's Run, dir. Michael Anderson (MGM, 1976).

The Matrix, dir. Larry and Andy Wachowki (Warner Brothers, 1999).

Metropolis, dir. Fritz Lang (Universum Film A.G., 1926).

Robocop, dir. Paul Verhoeven (Skyvision, 1987).

The Shining, dir. Stanley Kubrick (Hawk Films/Warner Brothers, 1980).

Soylent Green, dir. Richard Fleischer (MGM, 1973).

The Terminator, dir. James Cameron (Cinema '84/Orion, 1984).

ZPG (Zero Population Growth), dir. Michael Campus (Sagittarius Productions, 1971).

Index